54 Advances in Biochemical Engineering Biotechnology

Managing Editor: T. Scheper

Springer-Verlag Berlin Heidelberg GmbH

Metabolic Engineering

Volume Editors: H. Sahm, C. Wandrey

With Contributions by
L. Eggeling, A. A. de Graaf, R. Krämer,
D. Weuster-Botz, W. Wiechert, B. Sonnleitner

With 51 Figures and 13 Tables

 Springer

ISBN 978-3-662-14853-2   ISBN 978-3-540-48526-1 (eBook)
DOI 10.1007/978-3-540-48526-1

Library of Congress Catalog Card Number 72-152360

© Springer-Verlag Berlin Heidelberg 1996
Originally published by Springer-Verlag Berlin Heidelberg New York in 1996
Softcover reprint of the hardcover 1st edition 1996

Typesetting: Macmillan India Ltd., Bangalore-25
SPIN: 10122595   02/3020 - 5 4 3 2 1 0 - Printed on acid-free paper

Managing Editor

Professor Dr. T. Scheper
Institute of Technical Chemistry, University of Hannover
Callinstraße 3, D - 30167 Hannover, Germany

Managing Editor for this Volume:

Professor Dr. A. Fiechter
Institut für Biotechnologie,
Eidgenössische Technische Hochschule
ETH-Hönggerberg, CH-8093 Zürich, Switzerland

Editorial Board

Attention all "Enzyme Handbook" Users:

A file with the complete volume indexes Vols. 1 through 11 in delimited ASCII format is available for downloading at no charge from the Springer EARN mailbox. Delimited ASCII format can be imported into most databanks.

The file has been compressed using the popular shareware program "PKZIP" (Trademark of PKware INc., PKZIP is available from most BBS and shareware distributors).

This file distributed without any expressed or implied warranty.

To receive this file send an e-mail message to:
SVSERV@DHDSPRI6.BITNET.
The message must be: "GET/ENZHB/ENZ_HB.ZIP".

SPSERV is an automatic data distribution system. It responds to your message. The following commands are available:

HELP	returns a detailed instruction set for the use of SVSERV,
DIR (*name*)	returns a list of files available in the directory "name",
INDEX (*name*)	same as "DIR"
CD <*name*>	changes to directory "name",
SEND <*filename*>	invokes a message with the file "filename"
GET <*filename*>	same as "SEND".

Preface

Strain improvement is an essential part of process development for biotechnological products as a means of reducing costs by developing strains with increased productivity and yield the ability to use cheaper raw materials, or morespeialized desirable characteristics such as improved tolerance to high substrate and/or product concentrations. The procedure for the development of primary and secondary metabolites overproducing microorganisms so far has been mutagenesis and selection. Thus, the development of many highly productive industrial strains with this procedure has been largely an empirical process. The precise genetic and physiological changes resulting in increased overproduction of metabolites in many of these organisms have remained unknown. Success in attempts to further increase the productivities and yields of already highly productive strains will depend on the availability of detailed information on the metabolic pathways.

During the last few years, genetic engineering and amplification of relevant genes have become a fascinating alternative to mutageneses and random screening procedures. Application of recombinant DNA techniques to restructure metabolic networks can improve production of metabolites by redirected metabolite fluxes. However, as the metabolic activities of cell are accomplished by a network of more than 1000 enzymatic reactions and selective membrane transport systems, it is obvious that models and simulations are very important for carrying out effective metabolic design. Thus, in the first chapter techniques for the *in vivo* quantification of carbon fluxes and their control are presented. As the uptake of nutrients and the excretion of products are carrier mediated, transport processes which are also important steps for a complete describtion of the metabolic network, structural and functional properties of carrier systems from pro- and eukaryotic organisms are described in the second chapter. A throught understanding of the elements and mechanisms controlling the biosynthesis and transport of a metabolic should make it possible to influence its rate of overproduction in a predictable wav.

Furthermore, in the field of metabolic design, it is essential to combine improved knowledge about substrate uptake, metabolic networks and product excretion with improved biochemical engineering and modeling methods. Up till now, only a limited number of intracellular metabolites can be measured *in vivo*. NMR spectroscopy in principles is ideally suited as a non-invasive technique, however, it is not very sensitive. As described in the third chapter, a membrane-cyclone-reactor placed in the core of the magnet of the NMR spectroscope is quite useful. Thus a lot of intersting ^{13}P-NMR and ^{13}C-NMR data have been obtained from living cells. More quantitative information about metabolic fluxes can also be obtained from intracellular biopolymers with ^{13}C labeling experiments. Information on the metabolic network of microbial cells for example can be obtained by measuring the ^{13}C labeled amino acids after protein hydrolysis. From this information together with the improved knowledge about biochemical pathways it is possible to quantify a great number of fluxes in microbial cells as described in the fourth chapter. The data which may be obtained in future from the isotopomere signals may be sufficient to quantify hundreds of different fluxes. For new concepts of quantitative bioprocess research and development the interrelation between cell and environment must be studied in more detail. Better fundamental understanding results in process intensification and thus in a more stringent demand for high performance tools. Thus, improved online analysis methods for better bioprocess control and automation are described in the fifth chapter.

We hope that the reader of this book will enjoy some of the fascination we have experienced in a real interdisciplinary work. A biochemist who has to learn about adaptive parameter control in a bioreactor is in a demanding situation as a biochemical engineer who has to learn more about metabolic pathways. A NMR specialist has to learn about sterile technique and a mathematician about the different metabolic pathways. All this has helped us gain a better understanding of the fascinating field of metabolic flux analysis.

Jülich, December 1995   H. Sahm
                                                                                 C. Wandrey

Table of Contents

Quantifying and Directing Metabolite Flux: Application to Amino Acid Overproduction

L. Eggeling, H. Sahm, and A.A. de Graaf
Institute of Biotechnology, Research Center Jülich, D-52425 Jülich, Germany

The aim of understanding the metabolism of the cell in order to derive how high-level product formation can be achieved has led to astonishing progress with respect to the in vivo quantification of carbon fluxes and the quantification of flux control. This quantification usually holds only for one specific flux situation thus requiring additional approaches in order to obtain indications of how to produce a flux increase of relevance useful for overproduction. This is in part due to the complex responses of cells to high flux increases, including altered gene expression, alteration of energy metabolism, or accumulation of toxic intermediates, thus preventing a quantitative prediction of the targets to be altered. Instead, experimental techniques have to be used to examine the resulting consequences for each individual example. Currently, it is being found that branching points in biosynthesis, precursor supply in fueling reactions, and export of metabolites deserve particular attention in order to increase metabolite flux. By applying molecular techniques to such selected targets of the L-lysine and L-isoleucine biosynthesis of *Corynebacterium glutamicum*, high amino acid overproduction can be obtained with this bacterium.

Advances in Biochemical Engineering
Biotechnology, Vol. 54
Managing Editor: T. Scheper
© Springer-Verlag Berlin Heidelberg 1996

1 Introduction

It is a fact that knowledge of individual enzymes and pathways is insufficient to understand the entire metabolism. This view arises from such extreme poles as the theoretical considerations of the biochemical system theory [1, 2] and the practical example of overproduction of metabolites, where manipulating one reaction or even a set of reactions usually does not result in overproducers of industrial interest. Therefore, a comprehensive view of the entire cellular metabolism is mandatory. This is a current focus of research in biotechnology. In this field experts from a variety of disciplines, e.g. engineers, molecular biologists, biophysicists, biochemists, or physiologists, are busy using a variety of concepts and techniques. The aim is simply to quantify and direct the metabolite flux. The inherent aim is of course to predict in quantitative terms how to tailor the metabolism of the cell for maximal production of metabolites of industrial interest. As mentioned, the research to reach this goal comes from very different directions. For instance, the engineering approach originates from process engineering where control of the overall process is at the centre of attention and the cell is regarded merely as a catalytic component. This mechanistic view, focusing on extracellular parameters and using them as regulatory devices, has been extremely successful in improving and controlling production processes. The physiological approach is located at the other pole of research where the catalytic component itself, namely the cell, is at the centre. Here, for example, individual enzymes are deregulated, competing reactions reduced, or limiting activities increased. This approach is directed by and large towards qualitative physiological information. It has been most successfully applied to strain development. In fact, most of the classical strains producing amino acids, nucleotides, vitamins or antibiotics have been successfully bred without a complete understanding of the entire metabolic network. At this time, attempts are being made to combine these extreme poles and the diverse disciplines for a rational design of cellular metabolism with the aim of achieving maximum output of the desired metabolite.

It is the purpose of this review to present an overview of the different methodologies and concepts developed for flux quantification and their major areas of application. In particular, it will consider what kind of questions relevant for practical purposes can be answered by these approaches. Furthermore, it will describe how individual reactions or pathways structures in the cell can be identified. This appears trivial since the basic network of carbon flux seems to be known already from textbooks. However, there are exciting recent discoveries for organisms of long-standing use in industry. We then will outline an example of intracellular flux quantification using the most advanced methodology. Finally, the use and development of the practical means to direct intracellular fluxes will be described, together with their successful application.

2 Principles of Flux Estimations

Approaches giving access to cellular metabolism are shown in Table 1. Short general comments on their application and limitations are also given. Methods 1 and 2 are analytical tools based on theories and models. Methods 3–5 are experimental techniques for quantifying intracellular fluxes, and methods 6 and 7 are experimental techniques directed towards individual reactions.

2.1 Kinetic Based Models

The kinetic-based models for describing the metabolism rely on the kinetic parameters of each individual enzyme. The most detailed is the description of an enzyme reaction when the rate constants for the interconversion of the intermediate complexes (enzyme-substrate, enzyme-product) are also included in the respective formalism [3]. These reactions are simplified when specific rate constants are set to zero, for instance when the dissociation of the enzyme-substrate complex is considered to be irreversible. Then the description of that specific enzyme reaction becomes reduced and kinetic parameters can be obtained more easily from Lineweaver-Burk plots, Hill plots etc. [4]. Such kinetic parameters are the affinity constants for substrate(s), the affinity constants for activating or inhibiting effectors for the allosterically controlled enzymes, and the maximal velocity of the in vivo reaction. This information, together with that on concentrations of substrates and effectors, as well as fluxes through reactions, can be cast together in a model of the metabolism. Since, in vivo, the net fluxes through reactions are significantly smaller than the maximal velocities of enzymes, while the kinetic parameters hold for the entire activity range of the enzymes, this kind of analysis is in principle predictive. Thus, the advantage of such a description is that it follows a mathematical analysis for rate-limiting steps.

However, the obvious disadvantage is that for most systems the required detailed information will not be readily available. Another disadvantage is that there is no guarantee that all the expressions are comprehensive, simply because molecular research may not yet have identified all the specific reactions or regulatory properties that occur in vivo. The validity of the model depends, moreover, on the applicability of the enzyme constants that are derived from in vitro studies and therefore might be significantly different from the in vivo situation. Thus, in vivo, the polyelectrolytic conditions and protein concentrations are very different from the diluted in vitro conditions [5]. An additional problem is that an influence of concentrations on gene expression must be included, if known at all. Therefore, the requirements which have to be fulfilled usually prohibit the application of this approach for quantification of the entire cellular metabolism. Nevertheless, such a model has been made, for instance, for

Table 1. Means of analysing and quantifying fluxes

Type of analysis	Requirements	Advantage/Use	Disadvantage
(1) Kinetic-based models	Knowledge of the kinetic parameters of all reactions involved	Analysis of control architecture, and in theory prediction of limitations	The required information will usually be severely limited
(2) Control theories	Flux measurements at slightly altered enzyme activities	Numerical value for the importance of an enzyme. Values valid for one specific flux situation only	Experimentally difficult to achieve a set of minutely altered individual enzyme activities
(3) Tracer experiments	Specifically labeled substrates. Identification of label in products	Identification of pathway structure. Gives flux ratios at branch points	Yields relative fluxes
(4) Magnetization transfer	NMR-visibility Instrument necessary	The only direct in vivo quantification of reaction rates	Low sensitivity of NMR spectroscopy. Typically only rate constants between 0.05 and $5\,s^{-1}$ are accessible
(5) Metabolite balancing	Establishment of metabolite and of carbon balance	Readily accessible, yields absolute extracellular fluxes	Requires extreme simplification of intracellular network
(6) Enzyme analysis	Establishment of assay	Identification of pathways. Identification of allosteric control	No information on the contribution to in vivo use and flux
(7) Genetic analysis	Applicability of directed mutagenesis (reverse genetics)	Assay for the necessity of a reaction, for the presence of iso-enzymes	No information on in vivo use and flux. Gene deletion disturbs metabolism

the highly specialized red blood cell metabolism [6]. The model accounts for the pentose phosphate pathway, glycolysis, nuelcotide synthesis, transmembrane transport of key ions, pH dependence, magnesium complexation, electroneutrality, and osmotic balance. However, it has not been used as a predictive tool, nor have practical experiments been made to assay the validity of the model. For *Corynebacterium glutamicum* a kinetic-based model of the phosphoenolpyruvate-converting reactions was made [7] within the framework of a metabolite balance study. The kinetic representation used published kinetic constants for phosphoenolpyruvate carboxylase, pyruvate kinase, and five move enzymes, as well as estimates for intracellular metabolite concentrations and fluxes. Based on simulations of the kinetic representation, it was established that phosphoenolpyruvate carboxylase is of the utmost importance for high lysine yield. However, this conclusion was not experimentally verified. Instead, a subsequently constructed mutant devoid of phosphoenolpyruvate carboxylase was influenced neither in lysine overproduction for growth [8], which unambiguously negates the conclusion drawn from this kind of analysis.

2.2 Control Theories

Several extensive theories have been developed, not relying on enzyme kinetics, with the aim of quantifying the control strength of a particular reaction on the overall flux in a metabolic pathway. The theories are metabolic control analysis [1, 2], the biochemical systems theory [9] and a flux-oriented theory [10] combining aspects of the first two theories. All the theories include a form of sensitivity analysis, where the response of the whole system (usually the flux) to a small change (in the range of a few per cent) in a parameter (usually an enzyme activity) is quantified to derive specific control coefficients. Although inhibitors can be used to achieve the necessary small enzyme activity variations [11], or genetic instruments applied [12], there are severe practical problems in obtaining the sets of data required. Therefore dynamic approaches with in vitro systems are also used [13]. The current rapid sampling techniques developed to quantify intracellular metabolite concentrations [14, and see the contribution to this volume by Weuster-Botz et al.] will probably enable an easier data acquisition for flux analysis in the future. In contrast to the small changes required to derive control coefficients eventually, large enzyme activity variations can be obtained rather easily by genetic engineering. As a consequence, a theoretical framework has recently been presented by one of the originators of metabolic control analysis to analyze the response of metabolic systems to such large perturbations [15, 16]. Another recent development within metabolic control analysis is a "top-down approach", where the entire metabolism is divided into pathways which are connected by individual specific metabolites [17]. If this specific metabolite can be varied in its concentration, and the resulting flux changes quantified, information about the control exercised by complete sections of metabolism can be obtained, instead of for individual reactions only. Of major interest is, of course, how the different mathematical treatments of often abstract situations capture the relevant aspects of metabolism and control. In a "consumer test" [18] four different theories were compared to derive control coefficients for in vitro gluconeogenesis. Considerable difficulties were encountered in applying the theory to the practical situation. Slightly different answers were obtained to the question of which reactions exert significant control over pathway flux, but no answer could be obtained to the question of which regulatory mechanism is most important for pathway control under physiological conditions.

The important virtue of metabolic control analysis is that it has revealed that control of the overall flux of a metabolic pathway is distributed among several reaction steps instead of being localized at a single rate-determining reaction [19]. Apart from the practical difficulties mentioned before, a basic disadvantage of control analysis is that it only gives coefficients for one specific flux situation and that it has no predictive power for significantly different situations. Moreover, the assessment of control strength for allosterically controlled enzymes would pose even more severe practical problems for acquiring the necessary experimental data, although such allosterically controlled enzymes are of major importance when metabolite overproduction is attempted. Thus it seems in general more pracicable to oversynthesize the feedback-resistant enzyme in question by r-DNA techniques, thereby making an extreme change in enzyme activity, and simply evaluate the result. As an example, oversynthesis of feedback-resistant prephenate dehydratase within aromatic amino acid biosynthesis results in increased flux towards phenylalanine [21]. The goal of merely deriving a number for the control strength of the dehydratase reaction would have required knowledge of the actual in vivo activity of the enzyme, together with that of cellular effector concentrations at various fluxes. However, the success of flux increase can be judged directly by metabolite accumulation, and then new reactions have to be considered directly in the next step of flux increase. This could be, in the simplest case, a new "limiting" enzyme in the pathway, or even an entirely different kind of reaction. One such documented consequence of enzyme overexpression is the accumulation of high levels of pathway-related intermediates disturbing the cellular metabolism. This is presumably the case with many strains derived for metabolite overproduction, since impaired growth is often reported upon oversynthesis of controlling enzymes. As an example, the accumulation of toxic intermediates has actually been shown to occur in strains overproducing tryptophan [21], and in strains overproducing threonine [22].

2.3 Tracer Experiments

An important experimental technique for flux quantification is the use of tracers that can be either radioactive or stable isotopes. They are used on the assumption that they are biologically indistinguishable from their normal analogues. Although the main use of tracers is to analyze the structure of pathways, with a recent example of citrate metabolism in anaerobes [23], determination of the fluxes through pathways is also possible. Extensive use has been made of ^{14}C labeled precursors [24]. Specifically, [1-^{14}C]glucose and [6-^{14}C]glucose were used to quantify the use of the pentose phosphate pathway and glycolysis [25]. In an attempt to quantify cellular fluxes in E. coli, accumulation ratios of ^{14}C label into several cellular fractions were quantified. Together with a detailed analysis of ^{13}C label incorporated into glutamate, and a series of mass conservation equations, the flux rates through the tricarboxylic acid and dicarboxylic acid cycle were quantified [26, 27].

Of special benefit is the use of the ^{13}C isotope instead of the ^{14}C isotope. There are many reasons for this. First of all, incorporation of label in the individual carbons of one metabolite is directly accessible with ^{13}C NMR spectroscopy. This important information can thus be obtained more rapidly than with the ^{14}C isotope, where a chemical degradation would be required. As a consequence, a very large number of ^{13}C NMR studies of metabolic processes have been conducted. Using NMR, the incorporation of ^{13}C label in metabolites can even be followed in vivo, assuming that a high intracellular concentration of the metabolites exists [28, 29, and the contribution by Weuster-Botz et al. to this volume]. While the perspective of directly monitoring dynamic changes within the living cell is certainly exciting, the information retrieved from such studies has been limited and was largely of a qualitative nature due to the very limited set of metabolites accessible in vivo. Very useful information which can be derived from use of the ^{13}C isotope is the differentiation between isotopomers. These are the isotopically different physical species of one chemical compound. The mathematical treatment of isotopomer analysis is given in the fourth contribution to this volume (Wiechert et al.). This kind of analysis gives a very detailed view of cellular fluxes, enabling the quantification of back fluxes, futile cycles, or ordered transfer. The most impressive examples to date have been the studies with whole organs. The analysis of isotopomers of glutamate, glutamine, aspartate and γ-aminobutyric acid obtained from animal brain unequivocally showed that amino acid metabolism takes place in two different compartments [30, 31]. Moreover, the relative fluxes of pyruvate carboxylase and pyruvate dehydrogenase could be quantified for both compartments. In another example it was possible to distinguish between the use of three different sources of labeled acetylCoA entering the tricarboxylic acid cycle merely from the isotopomer analysis of succinate [32].

Although less spectacular, NMR studies have also been performed using the ^{2}H [33] and ^{15}N [34, 35] isotopes. The latter isotope, though very insensitive, was shown to be well-suited for quantification of nitrogen assimilation fluxes in competing enzyme reactions. Using $^{13}C/^{2}H$ multiply labeled glucose, the relative activities of the pentose phosphate pathway and glycolysis could be determined from a single incubation [36] by NMR.

2.4 Magnetization Transfer

The only direct method of quantifying in vivo flux rates is NMR magnetization transfer whose use, however, is limited because of the inherent insensitivity of NMR and because it requires the spin-lattice relaxation rates of the reactants and products to be of the same order as the reaction rate constants [36]. Furthermore, due to the complicated data analysis, only relatively simple unbranched reaction sequences can be studied in vivo. A well-known, extensively studied example is the creatine kinase reaction in muscle [37 and references therein]. Applications to biotechnologically relevant organisms are very rare. Only recently a magnetization transfer study was conducted to characterize the diffusion of ethanol over the cell wall of Zymomonas mobilis [38]. The measured diffusion rates were so high that even during the very rapid of glucose in this organism an intracellular accumulation of degradation the product ethanol is extremely unlikely to occur.

2.5 Metabolite Balancing

Metabolite balancing has been introduced by several authors, and extensively consolidated in a recent example for *Corynebacterium glutamicum* [39, 41, 44]. This technique assumes stationary conditions. Substrate consumption and product formation rates are determined usually comprising not more than ten estimates (substrate, product, side products, oxygen, carbon dioxide, ammonium, cell material). This yields the data directly for the respective in vivo transport reactions. In addition, with the known demands of central metabolites for cellular growth (oxaloacetate, pyruvate, α-ketoglutarate), this places several linear constraints on the use and rates of the basic metabolic pathways present, which of course have to be known from biochemical analyses. This kind of analysis can be of use for specific purposes, when only specific pathways are considered [42, 43] or if the contribution of flux for cellular growth is weak, as with hybridoma cells [44] and entire organs [45]. However, in the more complex bacterial systems, where product formation often occurs together with growth, and consequently use of the entire set of metabolic pathways is required, the large number of branching points usually prevents a detailed analysis [46].

Therefore, although the establishment of the carbon balance required is a straightforward approach to quantify the extracellular fluxes, a number of assumptions and simplifications concerning the entire cellular activities have to be included to derive intracellular fluxes. This includes the lumping together of reactions to simplify the network, which unfortunately often concerns the most interesting reactions. Furthermore, assumptions on the energy balance must be made, thus leading to further uncertainties, since in vivo knowledge of the number of coupling sites, or energy-dissipating reactions is often absent. Altogether, these approaches formally reduce the number of unknown variables in the network equations so that a mathematical solution is possible. Although metabolite balancing in itself does not result in a true detailed flux quantification, it can be used for the calculation of maximal yields of the metabolite of interest [43]. The technique is not predictive. Instead it is a snapshot of a particular flux situation, requiring additional approaches to recognize possibly limiting reactions [43, 47]. Of these, perturbations introduced by mutations and quantification of the resulting flux changes form one possibility. Another possibility is to perturb metabolism by specific inhibitors, addition of precursors, or sudden substrate increase. An example of how this can be achieved experimentally is given in the third contribution (Weuster-Botz et al.) to this volume.

3 Fluxes in Amino Acid Producers

The basis of any analysis or quantification of flux rates is of course a sound knowledge of the pathway structure existing in the producer interest, together with the mechanisms of intracellular flux control developed by that particular microorganism. Cellular metabolism can be divided into import reactions for supply with substrates, fueling reactions, biosynthetic reactions, polymerization and assembly reactions, and excretion reactions delivering the metabolite of interest and other products (Fig. 1). When metabolite overproduction is the subject of interest, it is evident that polymerization and assembly reactions do not have to be considered for identification of the flux-carrying reactions. Although major reaction pathways, as established, for example, in *Escherichia coli*, are also features of other microorganisms of industrial interest, the particular producer may have characteristic flux situations. This will be outlined below, taking *Corynebacterium glutamicum* with its subspecies *flavum* and *lactofermentum*, formerly called *Brevibacterium* sp. [48], as an example.

Since the discovery of this bacterium in 1957 [49] its glutamate excretion capability has been a stimulus for metabolite overproduction using microorganisms in general [50, 51], for the bioindustry [52], and, only recently, for the quantification of absolute intracellular flux rates [39]. Of the one million tons of amino acids produced annually, about 500 000 tons of glutamate and 200 000 tons of lysine are produced with *C. glutamicum*, and methods for the production of other amino acids are also exploited [22, 50, 53, 54]. *E. coli* is also used for amino acid production [55–57]. Other candidates for overproduction are *Serratia marcescenes* [58], *Bacillus subtilis*, and methylotrophic bacteria [59, 60]. *C. glutamicum* is a gram-positive actinomycete [61] isolated from soil [49], which is in accord with its resistance to starvation [62]. Specific features of this organism are: (i) it has a long tradition in biotechnology, and therefore knowledge of its process behavior is considerable, (ii) a deleterious accumulation of acetic acid as reported for *E. coli* [63–65] is not known; (iii) in contrast to *Bacillus subtilis* or *E. coli*, isoenzymes are less abundant in this bacterium, which might be due to its comparatively small genome size of about 3082 kb [66, 67]; (iv) the control architecture of the biosynthetic reactions is simpler (see Fig. 3); (v) control can easily be influenced [68, 69]; (vi) degradative enzymes for the two main bioproducts L-glutamate and L-lysine are unknown; and (vii) proteases are less abundant [70]. Due to the well developed genetics of this organism [71–73], *C. glutamicum* has recently been introduced as a system to oversynthesize and excrete foreign proteins [74, 75]. It is speculated that, probably due to less developed carbon flux control, *C. glutamicum* has a high capacity of energy dissipation [76]. However, since experimental data on the stoichiometry of the energy generation are scarce [77, 78], and consequences of possible imbalances speculative [20, 147], in the next section only examples of the identification of carbon-flux-carrying reactions are considered.

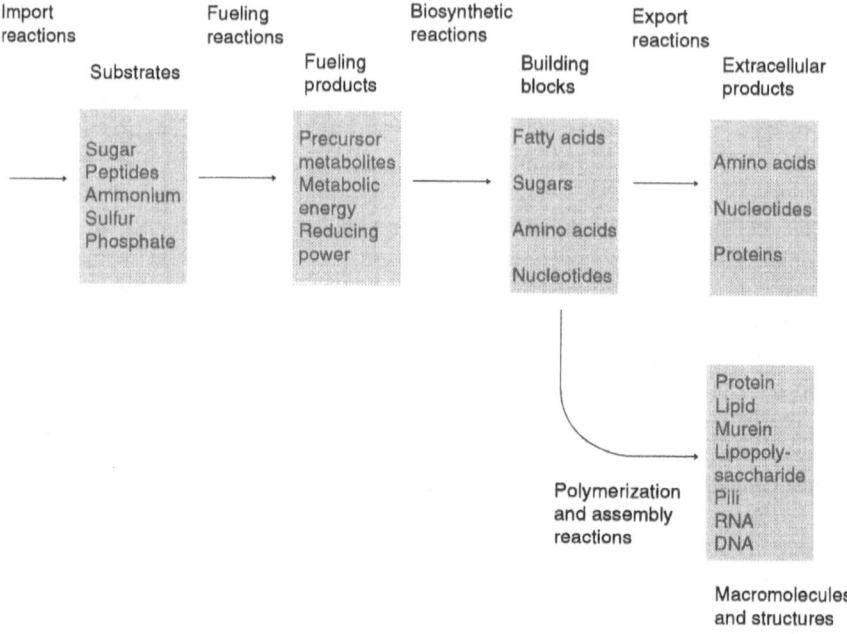

Fig. 1. Overview of metabolic reactions of relevance for the accumulation of extracellular low molecular weight metabolites

3.1 Identification of Flux-Carrying Reactions

As already mentioned, the transport reactions, the fueling reactions and the biosynthetic reactions have to be considered for metabolite overproduction. Although the identification of the corresponding reactions forms the basic prerequisite for any directed work on flux increase or modeling, even current knowledge of the relevant carbon-flux-carrying reactions is far from complete. The transport reactions will be described in the article in this volume by Krämer, and their relevance for lysine oversynthesis with *C. glutamicum* also in Sect. 4.3 of this article. The fueling reactions comprise an illustrative example of strain-specific pathway structures. The anaplerotic reactions link the precursor metabolites phosphoenolpyruvate (PEP), pyruvate and oxaloacetate (Fig. 2). Moreover they are linked via acetylCoA (derived from pyruvate) and oxaloacetate with the energy-generating tricarboxylic acid cycle, and, via the phosphotransferase system, simultaneously with sugar import. Therefore, the anaplerotic reactions function as a metabolic switchyard in connecting catabolism, anabolism, as well as energy supply. The sensitive control of traffic at this switchyard is also evident from the fact that in *E. coli* oversynthesis of PEP carboxylase results in increased growth yields [79], whereas oversynthesis of PEP carboxykinase results in reduced growth yields accompanied by increased

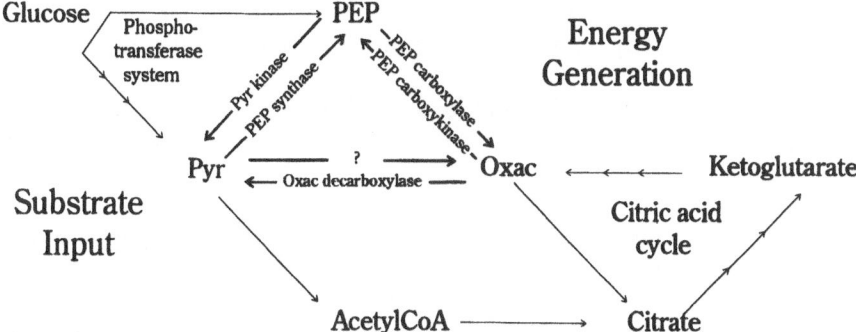

Fig. 2. Enzyme activities at the phosphoenolpyruvate node and relation of reactions to main cellular activities

glucose consumption and by-product formation [80]. In a mutant of the enterobacterium *Serratia marcescens*, oversynthesis of PEP carboxylase resulted in a 21% increase of threonine accumulation up to 63 g l^{-1} [81]. To solve the question of the in vivo use of the routes at this node an NMR isotopomer analysis would be particularly useful, as done for the rat [82]. In accord with this tight linkage of major subsets of metabolism is the recent discovery that one component of the phosphotransferase system is a trans-acting factor controlling expression of catabolic genes in gram-positive bacteria [83]. Although the metabolite flux at the switchyard is of central importance at an increased drain-off for overproduction, the structure of the corresponding reactions is not well understood for *C. glutamicum*. In this organism, at least seven enzymes have been characterized in vitro for the interconversion of phosphoenolpyruvate, pyruvate and oxaloacetate [84–86], with nine metabolites identified (acetyl-CoA, aspartate, ATP) affecting their activity [87]. Although the PEP carboxylase was considered to be of utmost importance for lysine overproduction [7, 50, 88], a mutant study with a producer devoid of PEP carboxylase activity revealed that the enzyme is dispensable for both growth and overproduction [8]. Also overexpression of the corresponding gene in a hyperproducer had no effect on final lysine yields. Recently, in an NMR-study using $^{13}CO_2$ and a PEP carboxylase mutant, it has been shown that another oxaloacetate-forming reaction must exist in *C. glutamicum* [86]. This molecular work is one step towards cutting the Gordian knot of the identification and use of the anaplerotic reactions in *C. glutamicum*. The simultaneous presence of two different enzymes catalyzing a forward and a backward flux for the interconversion of two metabolites is indicative of a sensitive regulation, since a small change in one of the two opposed activities has great consequences for the net flux. This *C. glutamium* example also aptly illustrates the specific features of enzyme analysis and genetic analysis (see Table 1). PEP carboxylase has on the one hand a high specific activity in vitro, but must not carry in vivo flux, and on the other hand the mutant of *C. glutamicum* devoid of PEP carboxylase certainly has no in vivo flux and must use a different reaction. However, no conclusion can be drawn from the mutant study on whether PEP carboxylase carries flux or not in the parent strain.

An example where enzyme and genetic analysis together with ^{13}C NMR analysis were used to their full potential for the elucidation of a pathway and its use is the diaminopimelate pathway of lysine synthesis in *C. glutamicum*. In this organism a split patway exists. Due to this uncommon feature of a biosynthetic reaction, the structure of the pathway for this economically important amino acid has in fact been a matter of considerable confusion [89, 90]. A simplified overview of this bacterial pathway of lysine synthesis, together with its connection to threonine, homoserine, and isoleucine biosynthesis is given in Fig. 3A. The pathway starts from the precursor metabolite aspartate derived from oxaloacetate which is provided by the tricarboxylic acid cycle and the anaplerotic reactions. After condensation with pyruvate, piperideine dicarboxylic acid (PDA) is formed, whose further conversion to the branch point intermediate D, L-diaminopimelate varies between bacteria. PDA can be converted via succinylated intermediates in a four-step reaction to D, L-diaminopimelate as in *E. coli* [91], or via acetylated intermediates as in *B. subtilis* [92]. But a one-step reaction is also known, i.e. in *Bacillus sphaericus*, where PDA is converted directly by an ammonium-incorporating dehydrogenase activity [93]. Surprisingly, activity determinations in *C. glutamicum* revealed the presence of the four-step succinylase variant together with the one-step dehydrogenase variant [90, 145, 146]. Although the general structure was clarified, the enzyme acivities alone, and in particular the low activities of the four-step reaction, gave no answer to the question whether the four-step reaction carries flux in *C. glutamicum* or not. This was solved by gene-directed inactivation of the one-step reaction showing that, in its absence, growth and lysine production was still possible for *C. glutamicum* [94]. Since in principle these kind of analyses give no answer to in vivo use (Table 1), ^{13}C NMR analyses were used with specifically enriched substrates to determine the actual use of both pathway variants [95, 145]. Quantification of the fractional enrichments in precursor metabolites and lysine showed that flux occurs both via the one-step and the four-step reactions simultaneously. Most importantly, in batch culture this use is dependent on the cultivation time (Fig. 4). This dynamic change in relative contribution of flux-carrying reactions could not have been detected by either enzymatic or genetic analysis. The relative use of the two pathways is directed by the ammonium concentration of the culture [95]. We propose that this luxuriant equipment of two pathways operating in common has to do with the importance of D, L-diaminopimelate, which is not only the ultimate lysine precursor, but is also required for cell wall synthesis.

3.2 Quantification of In Vivo Flux Rates

Having flux directioning in overproducers in mind, a detailed picture of the absolute in vivo flux rates within the central metabolism is essential. This serves

Fig. 3A, B. Structure of flux-carrying reactions within the aspartate family of amino acids: **A** in *C. glutamicum*; **B**: in *E. coli*, and regulation of enzymes by allosteric control (*open symbols*), or expression control (*closed symbols*)

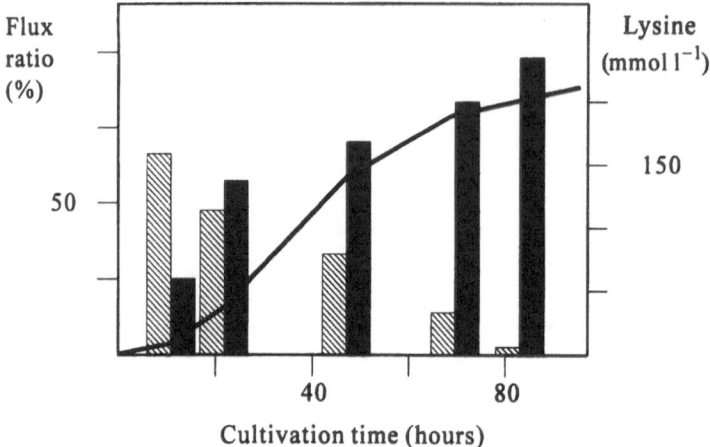

Fig. 4. The varying contribution of the one-step (*hatched*) and four-step variant (*filled*) of the diaminopimelate pathway for L-lysine synthesis with *C. glutamicum* in batch cultivation

Table 2. The special characteristics of the in vivo flux analysis for *Corynebacterium glutamicum*

1. Only data for metabolite and carbon atom balances are used. Energy considerations are not required
2. Specific enzyme activities and kinetic information on enzymes, as obtained from in vitro studies, are not required
3. With all fluxes no directionality is present
4. In selected cases fluxes are preset to be irreversible, but with arbitrarily chosen directionality
5. The conversion of citrate to α-ketoglutarate is not accompanied by rotation of the molecule
6. The conversion of α-ketoglutarate to oxaloacetate is accompanied by free rotation of the molecule

to analyze whether a reaction is active in vivo, to quantify in vivo fluxes through enzymes and pathways, to compare flux situations, and to provide a basis for strain improvement. However, in vivo flux quantification cannot simply be solved by one of the methods given in Table 1. Instead it requires the consistent combination of techniques and biochemical, physical, and mathematical expertise to achieve detailed and reliable flux estimations. This has been attempted in many studies with microorganisms [96–98], hybridomas cells [44, 102], and whole organs [99, 45], usually combining to various extents the means given in Table 1, but also using additional assumptions of energy stoichiometry and in vitro enzyme activities. In a recent approach to quantify the in vivo fluxes in *C. glutamicum*, the power of ^{13}C NMR analysis to follow individual carbons within the metabolism was merged with the simplicity of establishing carbon balances to a unifying comprehensive flux analysis [41]. This approach is distinguished by its very limited set of assumptions that relate only to the fate of carbon atoms in this organism (Table 2). An overview of the data generation and analysis is given in Fig. 5. A lysine overproducer of *C. glutamicum* was cultivated in continuous culture to yield directly the flux rates for substrate uptake, product release and biomass formation rates. For a detailed metabolite balance, the composition of the cell mass [41, 100] was taken into account. For instance, since cells contain 54 μmol tryptophan per g cell material, 54 μmol pentose phosphate, 54 μmol phosphoenolpyruvate and 54 μmol erythrose-4-phosphate are removed from the central metabolism for the formation of protein present in

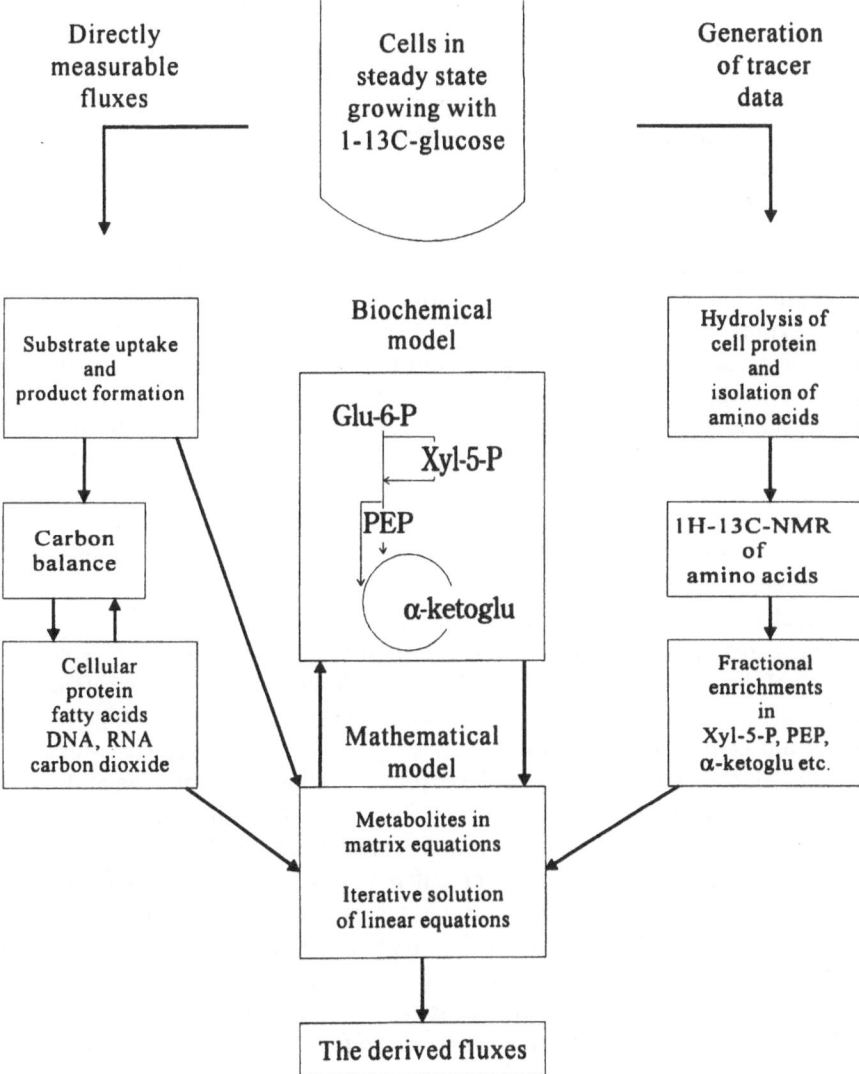

Fig. 5. Overview of the experimental data generation to quantify the steady-state in vivo fluxes in the cell

1 g cell mass. In this way, all intracellular fluxes removing precursor metabolites from the central metabolism for biomass formation were determined. As the second piece of information, the ^{13}C labeling of individual carbons within the precursor metabolites of the central metabolism was determined under steady-state conditions. For this purpose, the culture was grown with $[1-^{13}C]$glucose. The consequence is that in, metabolic and isotopic steady state, the fractional enrichments in the free precursor metabolites of the central metabolism are identical to those of the precursor metabolites fixed into cell material. Due to the known biochemistry of the biosynthetic pathways, the labeled carbon skeletons of the precursor metabolites are fixed in an exactly defined manner into cell material, e.g. pentose phosphates into nucleotides or the activated acetyl group into fatty acids. Since many different precursor metabolites are fixed into amino acids, as present in proteins, these are particularly informative for deriving label information on metabolites of the central metabolism. Consequently, amino acids were isolated from protein-hydrolysates of cells on a preparative scale and their fractional enrichments were determined by NMR spectroscopy, thus monitoring the ^{13}C labeling in 22 atoms of the precursor metabolites. The flux and labeling data were then processed using a newly developed numerical procedure. This highly approximate mathematical modeling using matrix calculus to express all metabolite and carbon balances, as well as the method for the iterative parameter fitting, is described in the fourth article of this volume (by Wiechert). Applying this comprehensive methodology, all major fluxes within the central metabolism of *C. glutamicum* were determined, such as the absolute fluxes through the pentose phosphate pathway, glycolysis, or the tricarboxylic acid cycle, thereby solving a long-standing question, i.e. the actual activity of the glyoxylate shunt, which was well determined under these conditions and is in fact very low (Fig. 6A). In addition to the very detailed quantitative estimation of all net metabolic fluxes, the high information content of the ^{13}C labeling data obtained from the NMR analysis also enabled the quantification of in vivo exchange reactions and backfluxes within the network. These are given in Fig. 12 of the contribution to this volume by Wiechert (page 140), where molar metabolite fluxes are given, instead of molar carbon fluxes as in Fig. 6A. An important conclusion from this determination is that exchange reactions are defined, which of course would not be of primary interest for applying molecular methods to achieve further flux increase in the situation analyzed.

As a further step ahead to derive general consequences of flux alterations in *C. glutamicum*, flux comparisons under different flux burdens are extremely informative [43, 101, 102]. It is clear that as few parameters as possible should be altered with respect to strain and cultivating conditions. Therefore an isogenic strain of the parent lysine overproducer was constructed [103], and biotin limitation used to produce glutamate instead of lysine. The same comprehensive methodology of intracellular flux analysis under steady-state conditions was applied as before. As can be seen from Fig. 6B, a massive restructuring of carbon flux is the result when the strain excretes glutamate instead of lysine. Thus, at an (incidentally) identical carbon flux for overproduction of either glutamate or

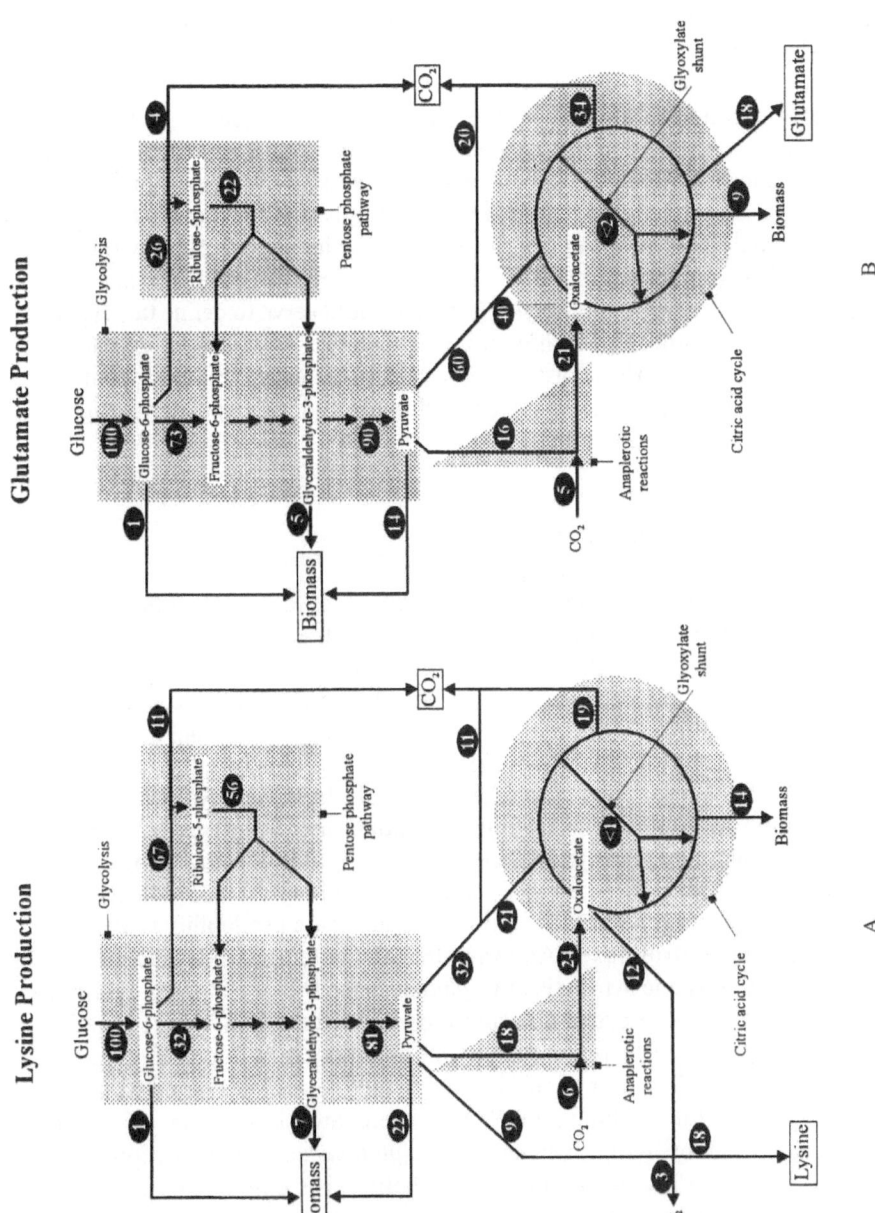

Fig. 6A, B. The in vivo carbon fluxes of *C. glutamicum*: **A** under L-lysine producing conditions; **B** under L-glutamate producing conditions. The numbers give the estimated net fluxes. They are expressed as percentage of the glucose uptake rate, which was 1.49 mmol g^{-1} dry weight h^{-1})

lysine, under glutamate-producing conditions a severely reduced flux through the pentose phosphate pathway occurs, but a strongly increased flux from pyruvate over the pyruvate dehydrogenase complex up to glutamate. It should be stressed that in this comparative analysis the strains analyzed differ only in the engineered aspartate kinase expression together with increased glutamate export activity. Therefore, in this example there is a direct correlation between induced export activity and resulting flux alterations. Consequently, in this case it is possible to conclude that in *C. glutamicum* energetic constraints might exist, since the reduced pentose phosphate pathway activity could be a result of increased isocitrate dehydrogenase activity supplying additional reducing power. This analysis shows the great flexibility of the fueling reactions within *C. glutamicum*. When more comparative flux analyses are available, trends in the observed changes will probably be visible, which could serve to define targets for the application of molecular techniques. For instance, it would be extremely interesting to examine the cause of molecularly achieved reduced pentose phosphate pathway activity or an increased flux through pyruvate dehydrogenase with respect to glutamate flux.

3.3 Mechanism of Flux Control

The in vivo flux determination described is a very detailed quantification of fluxes existing under the studied conditions. To direct fluxes actively, the mechanisms where and how fluxes are controlled must be known. The points of cellular flux control are destined by the pathway structure. Thus, control is usually primarily exerted at the entrance to pathways and at branching points. This can be seen from Fig. 3A, where the flux control points by either allosteric or repression mechanisms are given for the highly branched pathway of the aspartate family of amino acids within *C. glutamicum*. Surprisingly, there is currently no such type of lysine-dependent regulation known at the diaminopimelate distribution point in *C. glutamicum* [104]. In addition to flux control by allosteric mechanisms or repression, substrate availability is another cellular means to control fluxes. An example is the ammonium-dependent flux distribution within lysine synthesis of *C. glutamicum* [95], or the direct relation of lysine flux rate to the cytosolic aspartate concentration [105]. In Fig. 3B control of flux within the aspartate family of amino acids in *E. coli* is given for comparison. It is evident that in this organism flux control is exerted at many additional points. For instance, in *E. coli*, there are six enzymes of lysine synthesis repressed by lysine, whereas in *C. glutamicum* no such repression occurs [106]. Allosteric control in lysine synthesis exists only for the aspartate kinase [107], to shut down its activity in the presence of lysine plus threonine, whereas in *E. coli* the dihydrodipicolinate synthase is also regulated by this mechanism. The latter reaction competes with the homoserine dehydrogenase for aspartate semialdehyde. Although at first sight the control mechanisms used in *C. glutamicum* at this branch point by homoserine dehydrogenase repression and activity control seem to be sufficiently rigid for flux distribution, this is not the case, since repression of homoserine dehydrogenase by externally added methionine results in increased lysine flux [68].

4 Directing Metabolite Flux

Currently, quantitative predictions are not yet possible to direct cellular metabolite flux for overproduction purposes, but qualitative predictions can be made using the physiological information on the mechanisms of flux control described. In this respect the direction of intracellular metabolite flux by environmental control, to be described now, is an extreme technique, since although it is a quantitative technique, the cellular basis of successful flux directing is often ill defined.

4.1 Using Environmental Control

Process engineering is a powerful means to improve metabolite production. The simple goal of increasing product concentration requires increased supply of sugar, oxygen, ammonium, etc., and is merely a technical problem, whereas specific strategies are required when cellular metabolite flux is to be directed to increase product yield (product relative to substrate), or productivity (g product $l^{-1} h^{-1}$]. The basic principle is to assay whether certain values of easily accessible process parameters, like carbon dioxide evolution rate, oxygen uptake, growth characteristics or pH, correlate with peak activities in the desired yield performances. The strategy is then to try to hold these values of the extracellular parameters by environmental control, thereby extending the optimal flux situation within the cell [108, 109]. Illustrative examples are the peak activities of specific lysine [110] or isoleucine [54] productivity with *C. glutamicum* in batch culture correlating with a decreased growth rate. Therefore, growth limitation is successfully applied for several amino acid overproduction processes. Growth limitation can be attained by limiting potassium, phosphate [109], manganese [111], for instance, or by use of an auxotrophy [112], and is usually combined with feeding of sugar. The feeding strategy can be used to optimize for maximum molar yield or maximum volumetric productivity [110]. Only in specific cases can environmental flux control also be directly related to regulation of biosynthetic pathways. An example is lysine overproduction with a threonine-requiring strain of *C. glutamicum*. Since the presence of threonine inhibits the aspartate kinase activity, which controls the cellular flux into the lysine biosynthetic pathway (Fig. 3), the extracellular supply of threonine can be used to control the total metabolite flux towards extracellular lysine [110].

Although this engineering approach basically does not apply knowledge of cellular features, it apparently enables the critical optimal balance to be maintained between fueling reactions, biosynthetic reactions, and metabolite excretion. In the well-balanced situation, increased metabolite excretion might also be regarded as a vent to remove metabolites in the case of a surplus of fueling products. This is to some extent comparable to the overflow of central fueling products, exerted by several bacteria under limiting conditions at an excess of substrate [113]. Except for these kinds of interactions, the limitations applied will inadvertently also affect global regulatory control. An example is the well-known stringent response of enterobacteriacea, where deprivation of any amino acid, as often used in metabolite overproduction, triggers the regulation of numerous cellular activities [114]. Among these are, interestingly, a general stimulation of amino acid biosynthetic reactions. Obviously, intensive studies using molecular and biochemical work to elucidate the physiological background of limited growth must be correlated to application in industrial microbiology.

4.2 Altering Cellular Control

The cellular mechanisms to control the flux (see Sect. 3.3) also define the targets to increase the flux. The practical means to achieve this include (i) introduction of auxotrophy to overcome allosteric control, and for the same purpose (ii) use of feedback-resistant enzymes, and (iii) enzyme oversynthesis to overcome repression and increase the flux. At elevated flux and metabolite concentrations, however, many previously unimportant mechanisms might come into play, requiring strategies to be pursued to reach targets completely different from the original target of the strain construction. Thus, for instance, (iv) degrading activities of competing pathways might become relevant, requiring the inactivation of corresponding enzymes [115], (v) osmotic effects might occur [116–118], (vi) the tolerance towards intermediates might be reduced [22, 119], or (vii) excretion activities might become limiting [105].

One of the first steps in amino acid flux directing is to abolish allosteric control. For this purpose, strains that are auxotrophic for the regulatory metabolite can be used. However, due to the resulting increased process costs for starting material, it is generally preferred to alter allosterically controlled enzymes genetically so that they become insensitive to allosteric transition. This can be achieved by selecting strains that are resistant to an analog of the regulatory metabolite. Resulting strains will be altered in their allosteric transition of the enzyme in question [120], but other mutations might also cause the same phenotype like inhibited uptake of the analog [121] or its increased degradation [122]. Therefore, in vitro mutagenesis of the biosynthetic gene may be a more effective alternative than undirected mutagenesis [123]. An example is the mutagenesis of the deoxy-arabino-heptulosonate-phosphate synthase of aromatic amino acid synthesis of E. coli [124]. Chemical mutagenesis of the gene with subsequent selection of clones resistant to *p*-fluoro phenylalanine yielded enzymes with different degrees of feedback resistance. Another method, applied to feedback-regulated threonine dehydratase, used the toxicity of the threonine dehydratase reaction product ketobutyrate [125]. Clones of *E. coli* with deleterious growth after strong induction of the threonine dehydratase gene yielded a range of enzymes altered in their activity control. Interestingly, isoleucine inhibition is entirely abolished for the mutant Val-323-Ala enzyme, where the amino acid exchanged is located on the surface of the polypeptide (Fig. 7). This is in accord with the general view that the structural alterations of allosterically controlled enzymes are conditioned above all in the rotation of subunits with respect to each other in connection with slight transitions.

The degree of enzyme oversynthesis is another means of directing metabolite flux. To obtain high overexpression, it is standard practice to use multicopy plasmids. Promoters with high activity are also applied, or attenuation structures are deleted to increase transcript formation. Although a high overexpression might be successful for final product accumulation and predominates in

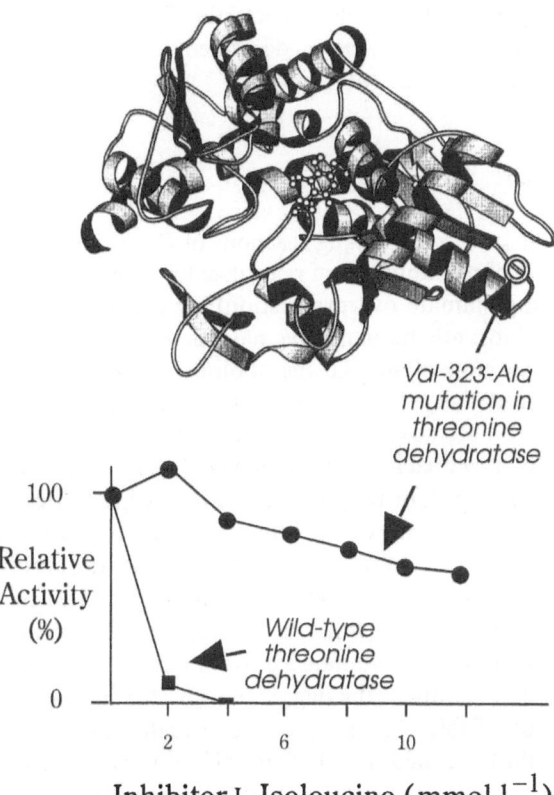

Fig. 7. Structure of the threonine dehydrase IlvA polypeptide (*top*), with location of the mutation resulting in feedback resistance to L-isoleucine inhibition (*below*)

current strain constructions, one critical issue is that high overexpression can have dramatic intracellular consequences. This is true in particular of feedback-controlled enzymes where the use of a feedback-resistant mutant enzyme leads to a massive restructuring of metabolite flux [105]. This can be macroscopically detectable as impaired growth, or detection of plasmid instabilities [69, 126, 127]. An example of massive flux alterations is aromatic amino acid synthesis in *C. glutamicum*. Depending on the overexpressed gene, the toxic intermediates indol [21], chorismate, or anthranilate [119] accumulate, accompanied by plasmid instabilities. A negative effect of high gene overexpression known for *E. coli* is a cumulative breakdown of rRNAs, and also the induction of heat shock proteins [128]. A negative effect on final yield with a plasmid-encoded biosynthetic enzyme activity is also known [129]. Consequently, for a smoother flux increase within a biosynthetic reaction sequence, it might be advantageous to apply lower overexpression, and in particular to overexpress a combination of biosynthetic enzymes. Low level expression is of particular use for flux distribution at branching points. An example is the reduced expression of dihydrodipicolinate synthase to reduce flux towards lysine, with resulting increased flux towards threonine accumulation [130]. Since affecting one branch point enzyme implicitly also affects the flux at the other, cumulative regulatory effects might occur, as is described for the isocitrate lyase and isocitrate dehydrogenase of *E. coli* [131].

4.3 Flux Increase Towards L-Lysine

Whereas within the network of fueling reactions of *C. glutamicum* it is surprisingly unclear what kinds of reactions are of relevance for lysine overproduction (as is also the case with glutamate overproduction), several targets within the more clearly structured biosynthetic pathways have now been identified.

The first enzyme controlling the metabolite flux in lysine biosynthesis is aspartate kinase. Its activity is controlled by the common presence of lysine plus threonine [107, 132]. Removal of aspartate kinase control in *C. glutamicum* either by the use of a feedback-resistant enzyme [120], oversynthesis of a feedback-resistant enzyme [69], or threonine auxotrophy [133] is essential to result in increased flux. Thus control of the kinase activity is of utmost importance. This is also concluded from the reverse approach, where the chromosomal copy of feedback-resistant kinase was replaced in a hyperproducer by the wild-type kinase [103]. In contrast to 220 mmol l^{-1} lysine accumulated by the parental strain, the isogenic mutant accumulated only 10 mmol l^{-1} lysine. However, the cellular background determines the degree of total flux increase towards lysine. As can be seen from Table 3, plasmid-encoded high feedback-resistant kinase activity results in the accumulation of 38 mmol l^{-1} lysine in the wild type, in the classically obtained lysine producer DG52-5 in 48 mmol l^{-1} accumulation, but does not affect the high lysine accumulation in the classically obtained hyperproducer MH20-22B. This illustrates that one chromosomal copy of feedback-resistant kinase can be sufficient to reach the high concentrations obtained and that other mutations must exist within the total network of reactions differentiating the three strains given in Table 3. In two *C. glutamicum* strains of different origin, however, an increase of up to 21% of lysine yield by one additional copy or plasmid-encoded feedback-resistant kinase was achieved at a maximum yield of 160 mmol l^{-1} lysine [129, 134].

The second reaction of relevance for flux increase is dihydrodipicolinate synthase. Although, according to general features of flux control this branch point enzyme was a candidate for lysine flux increase, its importance has not been recognized until recently. This is probably due to the fact that biochemical analysis has shown that the enzyme is neither allosterically controlled nor repressed [106]. The synthase competes with the homoserine dehydrogenase for the common substrate aspartate semialdehyde and is controlled in its activity [135] and synthesis as well [136, 137]. Most importantly, synthase oversynthesis is the only reaction having a positive effect in all the strains investigated, including the wild type, as well as the hyperproducer MH20-22B (Table 3). The repression of the homoserine dehydrogenase gene *hom* together with *thrB*, encoding the homoserine kinase, the activity control of homoserine dehydrogenase and the constant dihydrodipicolinate synthase activity forces this branch point to be considered as a whole.

The extensive genetic and physiological work on enzyme activity and product increase with the three different strains was extended for the direct assay of intracellular consequences [105]. The quantification of intracellular aspartate and lysine concentrations monitor the intracellular precursor metabolite availability and the potential of the entire biosynthetic pathway. Although only differing in allosteric kinase control, the wild type and the overproducer MH20-22B have comparable intracellular aspartate and lysine concentrations, but differ markedly in their lysine excretion rate (upper two lines in Table 4). Consequently, the excretion process is altered in the hyperproducer [103, 138]. As mentioned above, plasmid-encoded feedback-resistant aspartate kinase results in only a weak lysine accumulation of 40 mmol l^{-1}. The analysis of the

Table 3. Consequences of enzyme oversynthesis on lysine accumulation in three different *C. glutamicum* strains. The corresponding genes are given in brackets

Oversynthesized enzyme	Lysine accumulated (mmol l^{-1}) with strain		
	Wild type	DG52-5	MH20-22B
Control	0	40	240
Phosphoenolpyruvate carboxylase (*ppc*)	0	48	235
Aspartate kinase (*lysCFBR*)	38	48	220
Aspartatesemialdehyde dehydrogenase (*asd*)	0	41	240
Dihydrodipicolinate synthase (*dapA*)	11	48	260
Dihydrodipicolinate reductase (*dapB*)	1	40	235
Diaminopimelate dehydrogenase (*ddh*)	0	39	180
Diaminopimelate decarboxylase (*lysA*)	0	40	240

intracellular situation, however, revealed surprising new information. Thus, with plasmid-encoded aspartate kinase (pJC33) a high input of aspartate flux into the biosynthetic pathway was forced, as was also the case with the regulatory subunit of the kinase (pJC31), since the cellular aspartate concentration becomes greatly reduced to 1 mmol l^{-1} (lower two lines in Table 4). However, most interestingly, intracellular lysine piles up to an extremely high concentration of 220 mmol l^{-1}. Therefore, the biosynthesis pathway is apparently no longer limiting. Instead, the excretion of lysine is a third target to achieve flux increase for overproduction. Only recently has mutational evidence been obtained that there is a specific lysine export carrier in *C. glutamicum*, transporting cytosolic lysine to the medium [68]. After the initial identification of such a flux-carrying reaction [139], access to the corresponding molecular basis will open up entirely new perspectives for lysine overproduction with this organism.

4.4 Flux Increase Towards L-Isoleucine

Flux directioning towards isoleucine is more difficult to achieve than towards lysine. This is due to the more complex pathway structure with an additional branching off towards methionine (Fig. 3A), and the fact that four enzymes also carry out reactions involved in valine and leucine synthesis. As a consequence, in contrast to lysine synthesis, the simple oversynthesis of kinase does not result in isoleucine formation, nor does this result with homoserine dehydrogenase oversynthesis [140]. Consequently, at the beginning of strain construction a careful tuning of the five enzymes regulated by inhibition and repression is already required. An ideal parent strain is of course a lysine hyperproducer, since in such a strain a high flux up to the common intermediate aspartate semialdehyde is guaranteed. In pioneering work, starting from a lysine producer, the successful restructuring of metabolite flux from lysine towards threonine was obtained [123]. However, plasmid instabilities were noted upon overexpression of feedback-resistant homoserine dehydrogenase (*hom*FBR). It turned out that in several *C. glutamicum* strains *hom*(FBR) could not even be introduced [127, 140]. This problem was solved by the integration of one or several copies of

Table 4. Monitoring lysine biosynthesis and excretion by intracellular measurements in growing cultures of *Corynebacterium glutamicum*

Strain	Cytosolic concentration (mmol l$^-$)		Lysine excretion rate (mmol g^{-1} h^{-1})	Accumulated Lysine (mmol l^{-1})
	Aspartate	Lysine		
MH20-22B	10	9	0.57	240
Wild type	10	10	0	0
Wild type pJC31	1	92	0.05	24
Wild type pJC33	1	220	0.12	38

hom(FBR) in the chromosome and the use of a low copy number vector [22]. As a physiological consequence, threonine accumulation was already achieved with one copy of *hom*(FBR), as well as the accumulation of the intermediate homoserine and the threonine degradation product glycine. With increasing *hom*(FBR) copy numbers the accumulations of metabolites increased, but not in a linear fashion (Fig. 8). This finding, together with the previously noted plasmid instabilities, indicated severe cellular consequences, and stimulated an assay for the intracellular metabolite concentrations. Extremely high concentrations close to 100 mmol l^{-1} threonine and homoserine are present, whereas in the parent strain the respective concentrations are below 10 mmol l^{-1} (Fig. 8). This is certainly the reason for the instabilities observed. Most importantly, with three *hom*(FBR) copies threonine was accumulated intracellularly up to 100 mmol l^{-1}, but externally only 65 mmol l^{-1} was found. Therefore, also for threonine production with *C. glutamicum*, export is apparently a new reaction to be considered, limiting the accumulation process. Further flux directing for threonine overproduction requires biochemical and molecular studies on its excretion.

Of course strains with high flux towards threonine are ideally suited to enable further flux directing towards isoleucine. For this purpose feedback-resistant mutant enzymes of the threonine dehydratase of *C. glutamicum* were generated [125]. The influence of four different *ilvA* alleles (threonine dehydratase genes) in vectors of high and low copy number on isoleucine accumulation was quantified in strains with feedback-resistant homoserine dehydrogenase [54]. The best strain is characterized by *ilvA*(FBR) with the double mutation H278R-L351S on a high copy number vector and three *hom*(FBR) copies. It accumulates

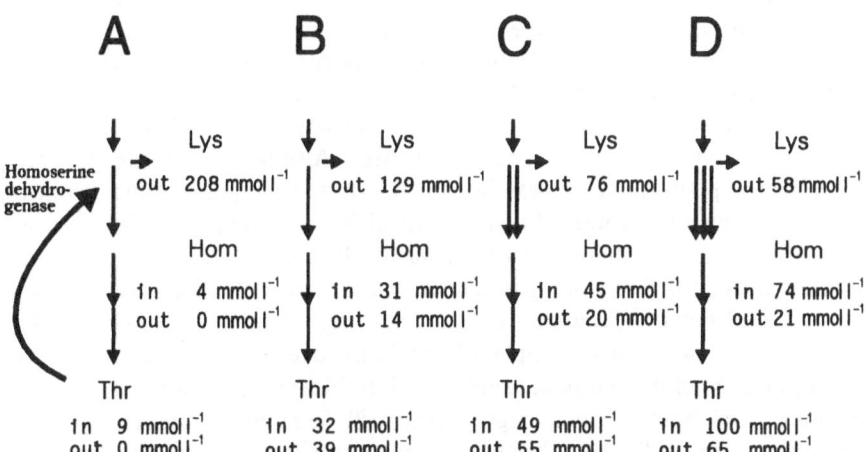

Fig. 8. Redistribution of flux between L-lysine and L-threonine synthesis as analyzed by the extracellular (out) and intracellular (in) accumulation of metabolites. A is the parent strain, B carries one, C two, and D three feedback-resistant homoserine dehydrogenase genes

96 mmol l^{-1} isoleucine without further accumulation of threonine, and has a highest specific isoleucine productivity of 0.052 g g^{-1} biomass h^{-1}. In this case high *ilvA*(FBR) overexpression is not only tolerated, but required for accumulation of the end product. The role of acetohydroxy acid synthase in the current flux situation has to be assayed, as well as isoleucine export. Due to the hydrophobic nature of isoleucine, this amino acid can permeate the cell membrane by diffusion [141], but it is also actively excreted [142, see also the contribution to this volume, by Krämer].

In an entirely different strain background with the wild type threonine dehydratase, a high flux increase from threonine to isoleucine was also obtained [143]. In this case a high *thrB* expression (homoserine kinase) was required, and achieved by IPTG induction, which can of course not be used in large-scale processes. However, this study shows that, within the long reaction sequence of isoleucine synthesis, comparable final isoleucine accumulations might be obtained either by an increased intracellular removal of threonine, as obtained by high threonine dehydratase activity, or by an increased threonine supply, as obtained by high homoserine kinase activity. Although current strains are already suited for isoleucine production, there is still a high potential for further flux increase by increasing further enzyme activities and careful adjustment of them as well.

5 Conclusions

Cellular metabolism, with its transport reactions, fueling reactions, biosynthetic reactions and assembly reactions can be described at very different levels by very different approaches. Each of these has its own advantages and success depends on the kind of meaures taken. If the aim is the purposeful design of metabolism, in the sense of predicting how to make or improve an overproducer, it is not as easy as saying "take this enzyme, improve its activity by a calculated factor, and you will obtain the desired product increase". Although there are now well consolidated platforms to assess in vivo fluxes and to quantify their control, there is no general theoretical concept useful for obtaining high flux increase, and it is unlikely that there ever will be due to the complex responses of the cell [144]. Instead, flux directioning is guided by knowledge of cellular activities and requires molecular techniques to assay for any qualitative prediction. The present studies of selected examples of metabolite overproduction have revealed the importance of branch points, entrance into biosynthetic pathways, and the export of metabolites, whose significance will have to be proven for other examples too. In addition to the present knowledge of flux controlling reactions, other mechanisms will certainly exist, since life is more than carbon flux, even for microorganisms.

6 References

1. Kacser H, Burns JA (1981) Genetics 97: 639
2. Heinrich R, Rapoport SM, Rapoport TA (1977) Progr Biophys 32: 1
3. Hayashi K, Sakamoto N (1986) Dynamic analysis of enzyme systems. Japan Scientific Societies, Tokyo
4. Segel LA (1975) Enzyme kinetics. John Wiley, New York
5. Cayley S, Lewis BA, Guttman HJ, Record JR (1991) J Mol Biol 222: 281
6. Joshi A, Palsson BO (1990) J Theor Biol 142: 69
7. Vallino JJ, Stephanopoulos G (1993) Biotech Bioeng 41: 633
8. Peters-Wendisch P, Eikmanns BJ, Thierbach G, Bachmann B, Sahm H (1993) FEMS Microbiol Lett 112: 269
9. Savageau MA, Sorribas A (1989) J Theoret Biol 154: 131
10. Crabtree B, Newsholme EA (1987) Biochem J 247: 113
11. Groen AK, Van der Meer R, Westerhoff HV, Wanders RJA, Akerboom TPM, Tager JM (1982) in: Sies H (ed) Metabolic Compartmentation, Academic Press, p 9
12. Flint HJ, Tateson RW, Barthhlemess IB, Porteous DJ, Donachie WD, Kacser H (1981) Biochem J 200: 231
13. Delgado J, Meruane J, Liao JC (1993) Biotechnol Bioeng 41: 1121
14. De Koning W, Van Dam K (1992) Anal Biochem 204: 118
15. Small JR, Kacser H (1993) Eur J Biochem 213: 613
16. Small JR, Kacser H (1993) Eur J Biochem 213: 625
17. Quant PA (1993) TIBS18: 26
18. Groen AK, Westerhoff HV (1990) in: Cornish-Bowden A, Cárdenas ML (eds) Control of metabolic processes. Plenum Press, New York, p 101
19. Kacser H, Acerenza L (1993) Eur J Biochem 216: 361
20. Neijssel OM, Teixeira de Mattos MJ (1994) Mol Microbiol 13: 179
21. Ikeda M, Nakanish K, Kino K, Katsumata R (1994) Biosci Biotech Biochem 58: 674
22. Reinscheid DJ, Kronemeyer W, Eggeling L, Eikmanns BJ, Sahm H (1994) Appl Environ Microbiol 60: 126
23. Ramos A, Jordan KN, Cogan TM, Santos H (1994) Appl Envion Microbiol 60: 1739
24. Blum JJ, Stein RB (1982) in: Goldberger R (ed) Biological Regulation and Development. Plenum Press, New York, p 99
25. Wood T (1985) in:Wood T (ed) The pentose phosphate pathway. Academic Press, Orlando etc, p 3
26. Walsh K, Koshland Jr. DE (1984) J Biol Chem 259: 9646
27. Walsh K, Koshland Jr. DE (1985) J Biol Chem 260: 8430
28. Den Hollander JA, Behar KL, Shulman RG (1981) Proc Natl Acad Sci USA 78: 2693
29. Bhaumik SR, Sonawat HM (1994) J Bact 176: 2172
30. Künnecke B, Cerdan S, Seelig J (1993) NMR in Biomed 6: 264
31. Lapidot A, Gopher A (1994) J Biol Chem 269: 27198
32. Jones JG, Sherry AD, Jeffrey FMH, Storey CJ, Malloy CR (1993) Biochemistry 32: 12240
33. London RE (1992) in: Berliner LJ, Reuben J (eds.) Biological Magnetic Resonance. Plenum Press, New York p. 6 (In Vivo Spectroscopy, vol 11)
34. Kanamori K, Weiss RL, Roberts JD (1987) J Biol Chem 262: 11038
35. Choi B, Roberts JE, Evans JNS, Roberts MF (1986) Biochemistry 25: 2243
36. Ross BD, Kingsley PB, Ben-Yosef O (1994) Biochem J 302: 31
37. Brindle KM (1988) Progr NMR Spectr 20: 257
38. Schoberth SM, Chapman BE, Kuchel PW, Wittig R, Grotendorst J, Jansen P, de Graaf AA (1994) Abstracts of the 12th European Experimental NMR Conference, Oulu, Finland p 43
39. Vallino JJ, Stephanopoulos G (1990) in: Sikdar SK, Bier M, Todd P (eds) Frontiers in bioprocessing, Boca Raton, CRC Press
40. Vallino JJ, Stephanopoulos G (1993) Biotechnol Bioeng 41: 633
41. Marx A, de Graaf AA, Wiechert W, Eggeling L, Sahm H (1996) Biotechnol Bioeng in press
42. Walker TE, Han CH, Kollman VH, London RE, Matwiyoff NA (1982) J Biol Chem 257: 1189
43. Stephanopoulos G, Vallino JJ (1991) Science 252: 1675

44. Mancuso A, Sharfstein ST, Tucker SN, Clark DS, Blanch HW (1994) Biotechnol Bioeng 44: 563
45. Kashiwaya Y, Sato K, Tshuchiya N, Thomas S, Fell DA, Veech RL, Passonneau JV (1994) J Biol Chem 269: 25502
46. Savinell JM, Palsson BO (1992) J Theor Biol 155: 201
47. Varma A, Palsson BO (1994) Bio/Technology 12: 994
48. Liebl W, Ehrmann M, Ludwig W, Schleifer KH (1991) Int J Syst Bacteriol 41255
49. Kinoshita S, Udaka S, Shimono M (1957) J Gen Appl Microbiol., Tokyo 3: 193
50. Kinoshita S, Nakayama K (1978) in: Primary products of metabolism, ed AH Rose
51. Furuya A, Sato A (1975) Appl Environ Microbiol 30: 480
52. Crueger W, Crueger A (1989) Biotechnologie, Lehrbuch der Angewandten Mikrobiologie. Oldenburg Verlag, München Wien
53. Eikmanns BJ, Eggeling L, Sahm H (1994) Antonie von Leeuwenhoek 64: 145
54. Morbach S, Sahm H, Eggeling L (1995) Appl Environ Microbiol 61: 4315
55. Patniak R, Spitzer RG, Liao JC (1995) Biotechnol Bioeng 46: 361
56. Kleemann A, Leuchtenberger W, Hoppe B, Tanner H (1985) in: Ullmanns' Encyclopedia of Industrial Chemistry 2A, 57-97. VCH, Verlagsgesellschaft mbH, Weinheim
57. Pátek M, Hochmannova J, Nesvera J (1993) Folia Microbiol 38: 355
58. Masuda M, Takamatsu S, Nishimura N, Komatsubara S, Tosa T (1993) Appl Biochem Biotech 37: 255
59. Motoyama H, Yano H, Ishino S, Anazawa H, Teshiba S (1994) Appl Microbiol Biotehnol 42: 67
60. Motoyama H, Anazawa H, Katsumata R, Araki K, Teshiba S (1993) Biosc. Biotech. Biochem 57: 82
61. Stackebrandt E, Woese CR (1981) in: Carlile MJ, Collins JF, Moseley BEB (eds) Molecular and cellular aspects of microbial evolution. Cambridge University Press, Cambridge
62. Tebbe C, Vahjen W, Munch JC, Feldman S, Sahm H, Gellissen G, Amore R, Hollenberg CP (1994) BioEngineering 10: 14
63. Dedhia NN, Hottiger T, Bailey JE (1994) Biotechnol Bioeng 44: 132
64. Aristidou AA, San K, Bennett GN (1994) Biotechnol Bioeng 44: 944
65. Hosono K, Kakuda H, Ichihara S (1995) Biosci Biotech Biochem 59: 256
66. Correia A, Martin JF, Castro JM (1994) Microbiology 140: 2841
67. Kalinowski J, Bathe B, Quast K, Pühler A (1995) Annual meeting of the Genetic society, Bielefeld, Sept. 18–21
68. Vrljic M, Konemeyer W, Sahm H, Eggeling L (1995) J Bacteriol 177: 4021
69. Cremer J, Eggeling L, Sahm H (1991) Appl Environ Microbiol 57: 1746
70. Liebl W, Sinskey AJ (1988) in: Ganesan AT, Hoch JA (eds) Genetics and biotechnology of bacilli, vol 2, Academic Press, p383
71. Wohlleben W, Muth G, Kalinowski J (1994) in: Rehm HJ, Reed G, Pühler A, Stadler P (eds) Biotechnology vol. 2, VCH, Weinheim, p457
72. Eikmannas BJ, Kleinertz E, Liebl W, Sahm H (1991) Gene 102: 93
73. Malumbres M, Gil JA, Martin JF (1993) Gene 134: 15–24
74. Billmann-Jacobe H, Hodgson ALM, Lightowerlers M, Wood PR, Radford AJ (1994) Appl Biochem Biotech 60: 1641
75. Billmann-Jacobe H, Wang L, Kortt A, Stewart D, Radford AJ (1995) Appl Environ Microbiol (1995) 61: 1610
76. Linton JD (1990) FEMS Microbiol Rev 75: 1
77. Shvinka YE, Viestur UE, Toma MK (1979) Microbiology-Engl Tr 48: 10
78. Kawahara Y, Tanaka T, Ikeda S, Sone N (1988) Agr Biol Chem Tokyo 52: 1979
79. Patniak R, Spitzer RG, Liao JC (1995) Biotechnol Bioeng 46: 361
80. Chao Y, Liao JC (1993) Appl Environ Microbiol 59: 4261
81. Sugita T, Komatsubara S (1989) Appl Microbiol Biotechnol 30: 290
82. Katz J, Wals P, Lee WP (1993) J Biol Chem 268: 25509
83. Hueck CJ, Hillen W (1995) Mol Microbiol 15: 395
84. Mori M, Shiio I (1987) Agr Biol Chem Tokyo 51: 129
85. Jetten MSM, Pitoc GA, Follettie MT, Sinskey AJ (1994) Appl Microbiol Biotechnol 41: 47
86. Peters-Wendisch P, Weidisch V, de Graaf AA, Eikmanns B, Sahm H (1996) Arch Microbiol submitted
87. Jetten MS, Pitoc GA, Follettie MT, Sinskey AJ (1994) Appl Microbiol Biotechnol 41: 47

88. Eikmanns BJ, Follettie MT, Griot MU, Sinskey AJ (1989) Mol Gen Genet 218: 330
89. Tosaka O, Takinami K (1978) Agr Biol Chem Tokyo 42: 95
90. Ishino S, Yamaguchi K, Shirahata K, Araki K (1984) Agr Biol Chem Tokyo 48: 2557
91. Kindler SH, Gilvarg C (1960) J Biol Chem 235: 3532
92. Weinberger S, Gilvarg C (1970) J Bacteriol 101: 323
93. White PJ (1983) J Gen Microbiol 129: 739
94. Schrumpf B, Schwarzer A, Kalinowski J, Pühler A, Eggeling L, Sahm H (1991) J Bacteriol 173: 4510
95. Sonntag K, Eggeling L, de Graaf AA, Sahm H (1993) Eur J Biochem 213: 1325
96. Jorgensen H, Nielsen J, Villadsen J, Mollgaard H (1995) Biotechnol Bioeng 46: 117
97. Patniak R, Spitzer RG, Liao Jc (1995) Biotechnol Bioeng 46: 361
98. Goel A, Ferrance J, Jeong J, Attai MM (1993) Biotechnol Bioeng 42: 686
99. Portais J, Schuster R, Merle M, Canoni P (1993) Eur J Biochem 217: 457
100. Neidhardt FC, Ingraham JL, Schaechter M (1990) Physiology of the bacterial cell. Sinauer Associates, Inc. Sunderland, Massachusetts
101. Ishino S, Shimomura-Nishimuta J, Yamaguchi K, Shirahata K, Araki K (1991) J Gen Appl Microbiol Tokyo 37: 157
102. Sharfstein ST, Tucker SN, Mancuso A, Blanch HW, Clark DS (1994) Biotech Bioeng 43: 1059
103. Schrumpf B, Eggeling L, Sahm H (1992) Appl Microbiol Biotechnol 37: 566
104. Oguiza JA, Malumbres M, Erian G, Pisabarro A, Mateos LM, Martin F, Martin JF (1993) J Bacteriol 175: 7356
105. Schrumpf B (1991) Diss. Universität Disseldorf
106. Cremer J, Treptow C, Eggeling L, Sahm H (1988) J Gen Microbiol 134: 3221
107. Shiio I, Miyajima R (1969) J Biol Chem Tokyo 65: 849
108. Satiawihardja B, Cail RG, Rogers PL (1993) Biotechnol Lett 15: 577
109. Coello N, Pan JG, Lebeault JM (1992) Appl Microbiol Biotechnol 38: 259
110. Kiss RD, Stephanopoulos G (1991) Biotechnol Prog 7: 501
111. Nara T, Misawa M, Kinoshita S (1968) Agr Biol Chem 32: 1153
112. Patek M, Krumbach K, Eggeling L, Sahn H (1994) Appl Environ Microbiol 60: 133
113. Tempest DW, Neijssel OM (1992) FEMS Microbiol Lett 100: 169
114. Cashel M, Rudd K (1987) in: Neidhardt FC (ed) Escherichia coli and Salmonella typhimurium. American Soc. of Microbiol p 1410
115. Aiba S, Tsunekawa H, Imanaka T (1982) Appl Environ Microbiol 43: 289
116. Gouesbet G, Blanco C, Hamelin J, Bernard T (1992) J Gen Microbiol 138: 959
117. Tomita K, Nakanish T, Kuratsu Y (1992) Biosci Biotech Biochem 56: 763
118. Booth IR, Higgins CF (1990) FEMS Microbiol Rev 75: 239
119. Katsumata R, Ikeda M (1993) Bio/technology 11: 921
120. Sano K, Shiio I (1970) J Gen Appl Microbiol Tokyo 16: 373
121. Seep-Feldhaus AH, Kalinowski J, Pühler A (1991) Mol Microbiol 5: 2995
122. Rossol I, Pühler A (1992) J Bacteriol 174: 2968
123. Katsumata R, Mizukami T, Kikuchi Y, Kino K (1986) in: Alaceviv M, Hranueli D, Toman Z (eds) Fifth Int. Symp on the Genetics of Ind Microorganisms, Zagreb, Yugoslavia, p 217
124. Ger Y, Chen S, Chiang H, Shiuan D (1994) J Biochem 116: 986
125. Möckel B, Eggeling L, Sahm H (1994) Mol Microbiol 13: 833
126. Ishida M, Kawashima H, Sato K, Hashiguchi K, Ito H, Enei H, Nakamori S (1994) Biosci Biotech Biochem 58: 768
127. Archer JAC, Solow-Cordero DE, Sinskey AJ (1991) Gene 107: 53
128. Dong H, Nilsson L, Kurland CG (1995) J Bacteriol 177: 1497
129. Jetten MSM, Follettie MT, Sinskey AJ (1995) Appl Microbiol Biotechnol 41: 76
130. Shiio I, Yokota A, Kawamura K (1989) Agr Bio Chem 53: 2169
131. LaPorte D, Walsh K, Koshland Jr DE (1984) J Biol Chem 259: 14068
132. Kalinowski J, Cremer J, Bachmann B, Eggeling L, Sahm H, Pühler A (1991) Mol Microbiol 5: 1197
133. Shiio I, Sano K (1969) J Gen Appl Microbiol Tokyo 15: 267
134. Lu J, Chen J, Liao C (1994) Biotechnol Lett 16: 449
135. Miyajima R, Otsuka S, Shiio I (1968) J Biochem Tokyo 63: 139
136. Miyajima R, Shiio I (1971) Agr Biol Chem 35: 424
137. Peoples OP, Liebl W, Bodis M, Maeng PJ, Follettie M, Archer JA, Sinskey AJ (1988) Mol Microbiol 2: 63

138. Bröer S, Eggeling L, Krämer R (1993) Appl Environ Microbiol 59: 316
139. Bröer S, Krämer R (1991) Eur J Biochem 202: 131
140. Eikmanns B, Metzger M, Reinscheid D, Kircher M, Sahm H (1991) Appl Microbiol Biotechnol 34: 617
141. Zittrich S, Krämer R (1994) J Bacteriol 176: 6892
142. Ebbighausen H, Weil B, Krämer R (1989) Appl Microbiol Biot 31: 184
143. Colon GE, Jetten MSM, Nguyen TT, Gubler ME, Follettie MT, Sinskey AJ, Stephanopoulos G (1994) Appl Environ Microbiol 61: 74
144. Bailey JE (1991) Science 252: 1668
145. Yamaguchi K, Ishino S, Araki K, Shirahata K (1986) Agr Biol Chem 50: 2453
146. Misono H, Ogasawara M, Nagasaki S (1986) Agr Biol Chem 50: 2729
147. De Hollander JA (1994) Appl Microbiol Biotechnol 42: 508

Analysis and Modeling of Substrate Uptake and Product Release by Prokaryotic and Eukaryotic Cells

Reinhard Krämer

Institute of Biotechnology, Research Center Jülich, D-52425 Jülich, Germany

Translocation of molecules and ions across cell membranes is an important step for a complete description of the metabolic network in terms of kinetics, energetics and control. With a few exceptions, most molecules cross the permeability barrier of the membrane with the aid of membrane-embedded carrier proteins. Uptake of nutrients (carbon, energy and nitrogen sources as well as supplements) and excretion of the majority of products are thus carrier-mediated transport processes. Consequently, they are characterized by particular kinetic properties of the respective carrier systems, they depend on energy sources (driving forces) which must be provided by the cell, and they are subject to regulation both on the level of activity and expression. They are thus fully integrated into the functional and regulatory networks of the cell. Structural (primary structure, conformation and topology) and functional properties (kinetics, energetics and regulation) of the different classes of carrier systems from both prokaryotic and eukaryotic membranes are summarized. The methodical requirements for a quantitative measurement of their function and possible pitfalls in transport studies are described, both for determination using isolated cells and for analysis in a bioreactor. The significance of transport reactions for biotechnological processes in general and for metabolic design in particular is discussed, with respect to nutrient uptake, product excretion and the occurrence of energy wasting combinations of transport reactions (futile cycles). Some examples are given where transport reactions have been incorporated into modeling approaches with respect to metabolic control, to flux analysis, to kinetic properties and to energetic demands.

Advances in Biochemical Engineering
Biotechnology, Vol. 54
Managing Editor: T. Scheper
© Springer-Verlag Berlin Heidelberg 1996

1 Introduction

In articles dealing with the analysis and improvement of metabolite production, in general two important aspects are considered. On the one hand, the particular growth conditions are discussed, including medium, supplements, substrate and oxygen supply, on the other hand the reactions within the cell are treated which refer to physiology, biochemistry and molecular biology of the enzymes involved in cellular metabolism, and their organization in terms of regulation. Frequently another aspect is more or less completely overlooked, i.e. that of membrane transport. Figure 1 makes clear that there is at least one, and in many processes two or more, steps involved in which molecules have to cross cell membranes. Cells need nutrients and supplements to allow metabolism, growth and reproduction, as well as production of certain metabolites. Moreover, the majority of biotechnologically relevant products is excreted into the medium by some kind of transport mechanism across the cell membrane. In processes in which the product is accumulated exclusively intracellularly (e.g. some vitamins or peptides), the second step is of course irrelevant.

Besides being of basic importance in functional models of metabolite production, the particular significance of transport reactions becomes obvious when the particular aspects, relevant in the context of this volume, are considered. As can be seen in Fig. 1, transport reactions may have significant impact on two other aspects, namely flux limitation and metabolic (carbon, nitrogen, energy, redox) balance. This is even more important in view of the fact that these reactions are, in general, tightly regulated by the cell, in order to maintain a defined metabolic steady-state and to avoid unwanted side reactions as well as futile cycles. Since the majority of the transmembrane fluxes need the input of metabolic energy in order to take place at relevant rates, the overall energy balance may also be significantly affected by the presence of transport events and in particular by their kind of mechanism and energy coupling.

Both knowledge on uptake and on excretion processes may be important when considering strategies to introduce or to improve overproduction of a certain product. There are a list of fundamental aspects in this respect, i.e. the simple availability of a particular uptake activity for a given substrate to be used, or the presence of an appropriate extrusion mechanism for a given metabolite to be produced. The entrance reaction, i.e the uptake of the major carbon source, may, on one hand simply be limiting by its maximum capacity, or it may be limiting under certain metabolic conditions due to regulatory properties with respect to its activity.

Besides these basic arguments concerning the strategy of using cells for overproducing certain compounds, the question why and under which metabolic conditions these kinds of transport mechanisms are present in the cell is of interest for a general strategy. This is of course not particularly relevant for uptake systems, in which usefulness of the cell is easy to understand. However, the presence of excretion systems, e.g. for amino acids, does not seem obvious.

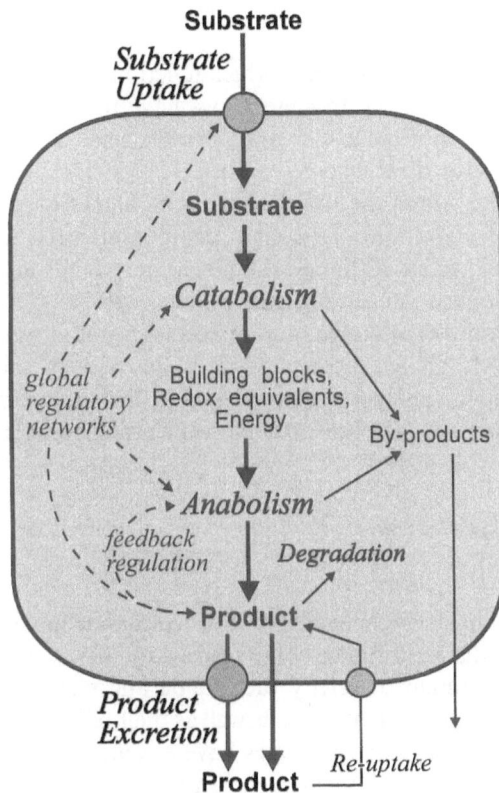

Fig. 1. Schematic representation of the significance of transport reactions among the principal metabolic steps in product formation by prokaryotic and eukaryotic cells. The overall carbon flux in general involves carrier-mediated substrate uptake as well as carrier-mediated or diffusion-controlled product release. In some cases, re-uptake of the product has to be considered

There are in fact two main lines of reasons for organisms massively overproducing certain metabolites [1]. On the one hand, overproduction of a certain metabolite can be caused by particular metabolic conditions of cells often connected to the lack of essential nutrients and consequently to a limitation of growth in the presence of an excess of carbon and energy (e.g. in the case of glutamate production in *Corynebacterium glutamicum*, of citrate production by several fungi, or overflow production of ethanol in yeast). This particular type of metabolic situation is called "overflow metabolism" (see below). On the other hand, cells with altered (mutationally changed) routes of anabolic and catabolic pathways are used for overproduction, thus leading to an increase in the particular end product and consequently to its excretion. Interestingly, these two very different approaches may be combined under particular circumstances, e.g. in the case of lysine production (see Sect. 4.2). Irrespective of the exact mechanism, it is interesting to know why these excretion systems are present in an organism under study. Improved knowledge of this aspect may lead to understanding of the particular type of regulation of these transport systems as well as to understanding of the particular metabolic situation which leads to overproduction and can thus, in principle, be used to optimize further the overproduction.

There are very different types of transport systems which function according to different types of mechanisms (see below). In general, we have to separate transport of (i) low-molecular-weight molecules such as carboxylic acids, amino acids and small peptides, mono- and disaccharides, fatty acids, nucleobases, nucleosides, or nucleotides, in other words the so-called building blocks of the cell constituents, from that of (ii) high-molecular-weight compounds such as proteins, polysaccharides or nucleic acids. Although the latter compounds are, of course, also of high biotechnological interest (at least proteins and polysaccharides), the transmembrane movement of these macromolecules is not as well understood as is the transport of smaller molecules. Thus, in view of the availability of a quantitative basis for analysis and modeling of the transport event, only transport of small compounds will be discussed in the present article.

2 Biochemistry of Transport Processes

In order to understand the physiological significance of transmembrane movement of molecules and ions and its importance with respect to the overall metabolic fluxes, it is helpful to summarize briefly the kinetic and energetic properties of membrane-embedded transport systems, as well as their relation to the regulatory networks in the cell. Transport processes have a central role in the cell metabolism. Except for a few hydrophobic (e.g. butanol) and very small (e.g. ethanol and glycerol) molecules and for gases, the plasma membrane is more or less impermeable for most molecules. Thus carrier systems in the cytoplasmic membrane are responsible for a variety of functions essential for the proper function of every cell. (i) They basically allow the passage of molecules across the otherwise impermeable phospholipid bilayer. (ii) When coupled to an appropriate source of energy, their activity may lead to substantial gradients (up to 10^6-fold) across the cytoplasmic membrane. This may be used either to accumulate an essential nutrient within the cell, or to export effectively unwanted end products or toxic compounds. (iii) Besides this, transport systems are involved in a variety of other functions, e.g. osmoregulation, pH and ion homeostasis, energy transduction, sensory processes and signal transduction.

Carrier systems are found in all biological membranes. In the case of prokaryotic cells, carrier proteins are embedded in the cytoplasmic membrane, and, in addition, some pore (channel) proteins are found in the outer membrane or cell wall. In eukaryotic cells the situation is more complicated. Both for uptake of certain nutrients and for excretion of products, transport across intracellular membranes of organelles has to be taken into account as well as the cytoplasmic membrane. For biotechnologically relevant processes involving small metabolites this refers almost exclusively to the mitochondrial membrane (e.g. citrate production in fungi).

2.1 Transport Proteins and Mechanisms: Carriers, Channels, Pumps

A basic knowledge of the different classes of transport systems is essential for the correct integration of the transport step into kinetic models and into appropriate energetic considerations. Because of the substantially greater amount of knowledge of uptake systems in comparison to efflux processes, mechanistic concepts have been developed almost exclusively for the uptake of molecules and ions. The basic energetic and kinetic principles of transport are of course identical for uptake and excretion. However, some of the mechanisms are not active in the efflux direction, e.g. binding protein-dependent transport systems and phospho*enol*pyruvate:sugar phosphotransferase systems. Furthermore, there are other systems which in general catalyze the vectorial reaction in one direction only, i.e. either to the outside or to the inside, e.g. some ATPases. It is now widely accepted that transport systems should be classified according to the concept introduced by Mitchell [2], based on the utilization of energy sources for transport. The different classes are briefly outlined here (Fig. 2) and the driving forces responsible for the vectorial process of translocation are summarized (Fig. 3). For a more detailed description the reader is referred to other reviews and books on this topic [1, 3–7].

(I) In simple, passive mechanisms, molecules cross the permeability barrier of the membrane without the involvement of a carrier protein ("simple diffusion"). Typical examples are the diffusion of small and/or hydrophobic molecules such as ethanol, propanol or acetone. The driving force for a passive, diffusion-controlled flux of a neutral molecule across the membrane is simply its chemical gradient (Fig. 3). In particular cases, translocation of a molecule may be facilitated by the presence of a pore protein. Movement of molecules and ions through the water-filled channel of such a protein may involve properties of different types of mechanisms, from "pore diffusion" to carrier-like mechanisms [1, 8, 9] (Fig. 2, I). Channel-mediated mechanisms of biotechnological significance in microorganisms are rare, examples being the passage of molecules through the pore proteins of the outer membrane of Gram-negative bacteria (see Sect. 3.3).

(II) The only driving force in secondary transport is the electrochemical energy of a molecule or ion (Fig. 2, II). This energy is utilized to drive the uphill transport of another molecule, i.e. against its own concentration gradient. This is achieved either by cotransport (symport) or countertransport (antiport) of the driving and the driven molecule or ion, depending on the direction of the electrochemical gradients. Because of the similarity in mechanism, carrier-mediated unidirectional transport simply driven by its own electrochemical gradient, also called "facilitated diffusion", is classified as secondary transport (uniport). Since symport and antiport may lead to uphill transport of the driven ion, they have also been classified as "secondary active". In secondary transport, the driving forces may include a chemical part and an electrical part (Fig. 3). The former includes the chemical gradients of the transport substrate S ($\Delta\mu_S/F$) and

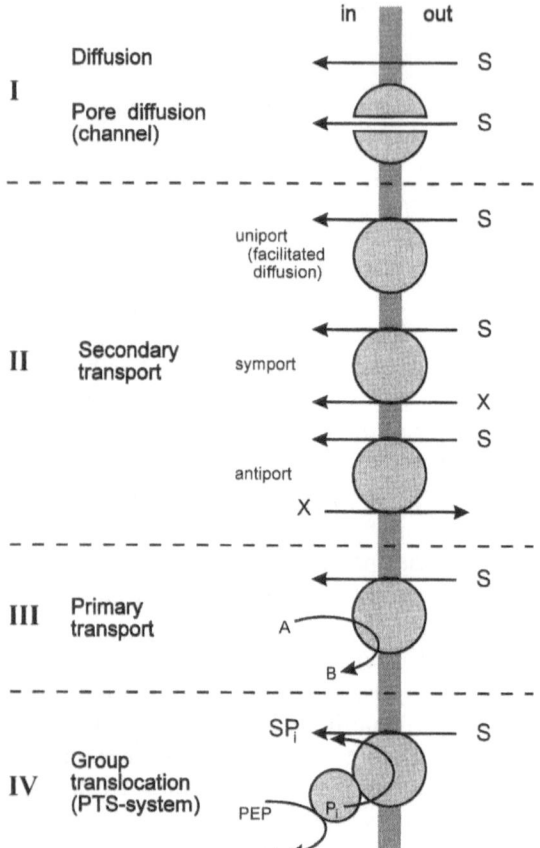

Fig. 2. The different classes of transport mechanisms, in terms of the utilization of energy sources. For further details see text

of the coupling ions X ($\Delta\mu_X/F$), which in the case of H^+-coupled secondary transport means $Z\,\Delta pH$ (F represents the Faraday constant and $Z = 2.3\,RT/F$). The electrical part includes the contribution of the membrane potential if the transport substrate and/or the coupling ions are charged. The different driving forces, as listed above, must be multiplied according to the transport stoichiometry and the number of individual charges (see Table 1). Molecules and ions with different charges, both in sign and number, as well as different numbers of molecules per carrier protein are transported. Since these variations have to be considered for uniport, symport and antiport, the list of possible contributions to various driving forces in different secondary transport systems is immense. Energy coupling in secondary transport in general and selected aspects of this field have been treated in numerous reviews and books [1, 3, 4, 7, 10–12]. Secondary systems are the most common transport systems and are frequently coupled to the transmembrane movement of Na^+ or protons. They are responsible for uptake of all nutrients in eukaryotic cells (e.g. carbohydrates, amino

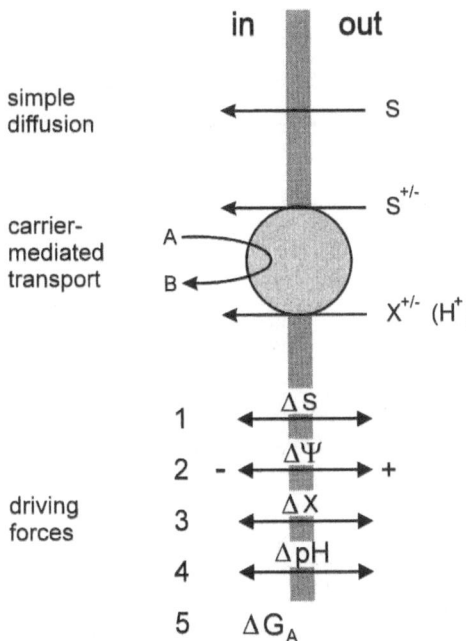

Fig. 3. Different driving forces influencing transfer of molecules and ions across membranes. The transfer rate of simple diffusion (*upper part*) and carrier-mediated transport (*middle part*) is modulated by different driving forces. These are the chemical gradient of the transport substrate (ΔS, actually $\Delta\mu S/F$, see text), the electrical potential or membrane potential ($\Delta\Psi$), the chemical potential of cotransport ions such as H^+ or Na^+ (ΔX), the pH gradient (ΔpH), and, if a primary system is involved, the free energy ΔG_A of the coupled chemical reaction

acids, carboxylic acids) and for a major part of nutrients in prokaryotic cells (e.g. amino acids, carboxylic acids, some carbohydrates), i.e. all except those transported by carriers of classes (III) and (IV).

(III) In primary transport the vectorial reaction of translocation is directly coupled to some kind of chemical or photochemical reaction. Primary transport systems thus convert light (e.g. photosynthetic electron transport) or chemical energy (e.g. binding protein-dependent systems and ion-translocating ATPases) directly into electrochemical energy, i.e. the electrochemical potential of a given molecule or ion (Fig. 2, III). In primary transport mechanisms the driving force is the free energy of the chemical or photochemical reaction, and in most of these systems the free energy is highly negative. This should lead to high electrochemical potentials and consequently to high concentration gradients of the transported substance. Because of inherent leaks in the protein/phospholipid membrane as well as slips (mechanistic imperfections) in the carrier protein itself, the driving forces in primary systems are often far from being in equilibrium with the electrochemical potential of the driven ion. Typical examples of this class of transport systems are the various kinds of ATPases, i.e. the ATP synthases (F-type) in prokaryotic plasma membranes and in mitochondrial membranes of eukaryotes, the plasma membrane ATPases (P-type) responsible for ion transport in eukaryotic (e.g. Na^+, K^+-ATPase) and prokaryotic cells, and finally the ATPases from vacuolar membranes (V-type). Other common examples are the carrier systems of the ABC-family (named after the structural motif of an

Table 1. Different driving forces in some secondary systems

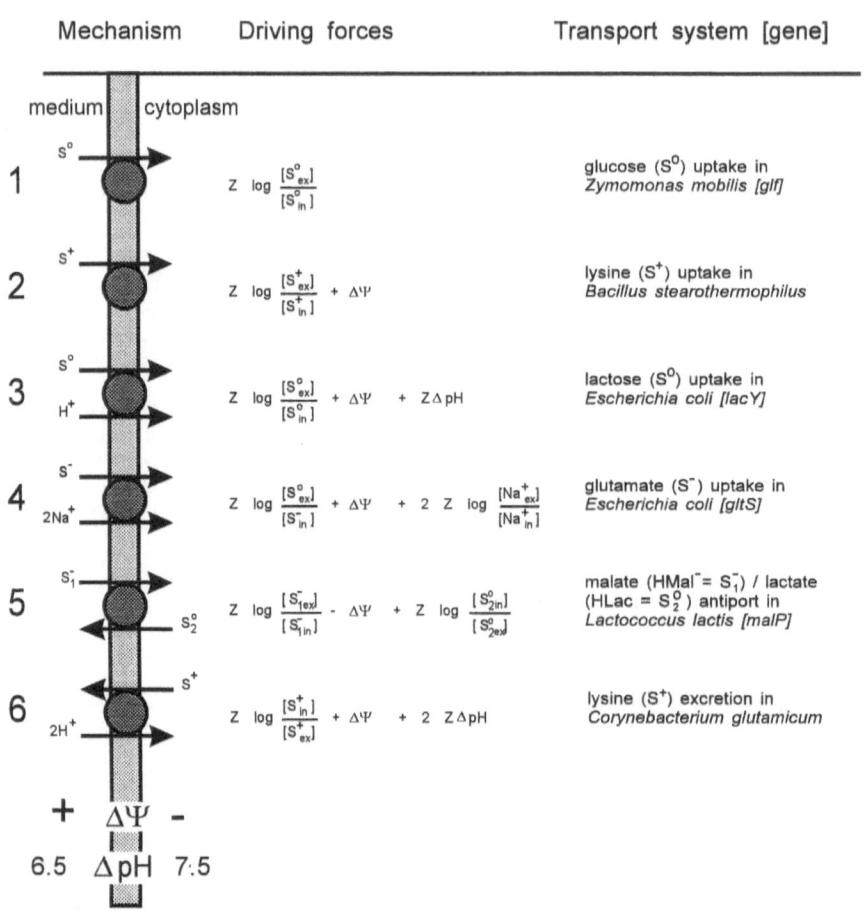

Examples 1 and 2 are uniports, 3 and 4 symports and no 5,6 antiports (lysine/2 OH⁻ symport [91] is equivalent to lysine/2 H⁺ antiport as shown in the figure). For reasons of simplification, the sign of the different driving forces was adjusted in the following way: positive sign, driving in the direction of the indicated transport system; negative sign, opposing force. (S = substrate, $Z = 2.3\,RT/F$, $\Delta\Psi$ = membrane potential, ΔpH = pH gradient)

ATP-binding cassette) which include the binding protein-dependent uptake systems of bacteria (accepting very different substances like amino acids and peptides, mono- and disaccharides and others), and the mutidrug resistance proteins known from all cells.

(IV) Group translocation differs from all other mechanisms by the fact that the transported molecule becomes chemically modified during translocation. The only known systems catalyzing group translocation are the phosphoenol-pyruvate:sugar phosphotransferase systems (PTS) in bacteria. From the energetic point of view, they are closely related to primary transport (Fig. 2, IV). These systems are responsible for the uptake of various sugars and sugar alcohols in prokaryotes.

2.2 Kinetics of Transport

When measuring transfer of molecules and ions across biological membranes, the dependence of the observed transport rate on the substrate differs characteristically, according to whether a diffusion-controlled process or carrier-mediated transport is studied (Fig. 4). The rate of diffusion increases linearly upon increasing the concentration of the transported molecule, whereas carrier-catalysed transport reaches a maximum value, i.e. the substrate-binding site of the transporter becomes saturated with its ligand, the transported molecule. Furthermore, in contrast to diffusion, carrier-catalysed transport depends on the concentration and activity of a proteinaceous compound of the membrane, namely the carrier protein, and can be influenced by the presence of other substances, e.g. inhibitors or activators (Fig. 4). Kinetics of carrier-mediated transport can be treated in perfect analogy to enzyme kinetics. In fact, a transporter can be regarded as an enzyme, the reaction of which is not a change of the substrate's structure but of its location. Graphical analysis of the experimental

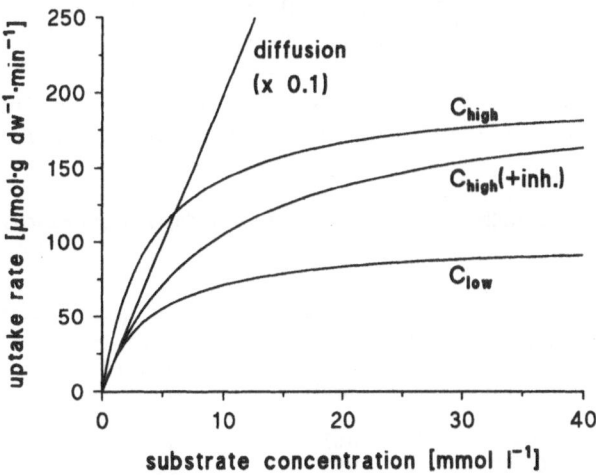

Fig. 4. Kinetics of carrier-mediated transported processes and diffusion-controlled membrane transfer. Uptake of D-glucose at 15 °C by the glucose facilitator from Z. mobilis, when heterogolously expressed in glucose-negative E. coli strains [80] at different levels (C_{low} and C_{high}), as well as in the presence of 10 mmol l^{-1} of the competitive inhibitor D-xylose (C_{high}(+ inh)). For comparison, acetate uptake by Corynebacterium glutamicum at pH 6.0 is shown (diffusion), where carrier-mediated acetate uptake is not active [142]. Since the rates of acetate diffusion are actually very high, for a better comparison of the shape of the various curves the values of the measured actate diffusion rates have been divided by a factor of ten.

data can also be performed, similar to that known from enzyme kinetics. In particular, this refers to inhibition kinetics. The well-known plots, such as velocity vs substrate concentration (Michaelis-Menten diagram, see Fig. 4), as well as reciprocal plots (e.g. Lineweaver-Burk diagram, Dixon plot) to identify the type of inhibition (competitive, noncompetitive etc.) can be used.

Basically, a molecule can cross the permeability barrier of the membrane by diffusion, which means movement of the transport substrate (S) without direct interaction with a membrane protein. The rate of unidirectional transfer (v_1) from one compartment (I) to the other (II) is directly proportional to the concentration of the molecule (S) in compartment I. A corresponding equation holds for the reverse flux from side II to side I. The net flux of a molecule (S) from side I to side II is then described by

$$V = PA([S]_I - [S]_{II}) \qquad (1)$$

where P is the permeability coefficient and A the area of the membrane [3]. Net transport by diffusion occurs only from a compartment of higher to that of lower concentration and is not specific. The transfer rates of diffusion-controlled processes are given in Table 2 for some relevant molecules. These examples demonstrate that, for most (hydrophilic) molecules, passive diffusion across the cytoplasmic membrane is low and not of physiological or biotechnological relevance. On the other hand, diffusion of urea, as an example, is significant, irrespective of the fact that, in addition, transport systems are found in many prokaryotic and eukaryotic organisms. The rate reported for phenylalanine in Table 2 is significant, at least at high internal phenylalanine concentrations, and may, for example, be sufficient for phenylalanine excretion by recombinant producer strains of E. coli and other bacteria [13, 14].

Carrier-mediated transport involves the participation of a proteinaceous component of the membrane, i.e. the transporter or carrier protein. It is not

Table 2. Comparison of diffusion-controlled and carrier-mediated solute fluxes across bacterial plasma membranes

Transported Solute	Typical transfer rate		Carrier-mediated (V_{max})
	Diffusion-controlled at a concentration difference of		
	$10\ \mu mol\,l^{-1}$	$10\ \mu mol\,l^{-1}$	
K$^+$	0.00002	0.02	100
glutamate	< 0.00005	< 0.05	25
glucose	0.001	1	50
isoleucine	0.0015	1.5	3
phenylalanine	0.008	8	1
urea	0.04	40	5

The transfer rates (in $\mu mol\,min^{-1}\,g^{-1}$ dry mass) for diffusion-controlled processes are calculated on the basis of the known passive permeability of bacterial plasma membranes, whereas the rates of carrier-mediated processes represent examples of typical transport systems specific to the respective solutes at full saturation (V_{max})

always that simple to prove unequivocally the involvement of a transport protein in a transport process. The major arguments indicating the presence of a transporter are mainly of a kinetic type. These are, similar to enzymes, (i) saturation kinetics, (ii) substrate specificity, (iii) inhibition by specific agents, and (iv) the particular kinetic observation of counterflow of substrate molecules between the two compartments (see Sect. 3.2).

For a basic kinetic discrimination between carrier-mediated transport and diffusion, it is very convenient to discuss secondary uniport, also called facilitated diffusion (see above). Due to their particular mechanism, other types of transport systems, e.g. primary mechanisms, are relatively easy to discriminate from diffusion-controlled transfer. Facilitated diffusion is a passive process, but since it is carrier-mediated, the net flux is not simply proportional to the concentration difference of the transported molecules across the membrane. Many transport systems, not only those of facilitated diffusion but also the majority of the transport mechanisms listed above, can phenomenologically be described by Michaelis-Menten kinetics.

$$v = V_{max} [S]/(K_M + [S]).$$ (2)

K_M is used by analogy to enzyme kinetics, it means the concentration of the transported molecule at which the transport rate v reaches half its maximum (V_{max}), and is also called K_t (for transport). Typical K_M values for transport systems vary from the submicromolar range, e.g. for some binding protein-dependent systems, to the millimolar range, e.g. for sugar transport in yeast. The observed variation in the capacity of transport systems (V_{max}) is at least as large. Transport systems for uptake of some growth supplements, e.g. vitamins or certain amino acids, are at the lower end (< 0.1 μmol g^{-1} dry weight min^{-1}), and some uptake systems for low energy-yielding substrates of anaerobic bacteria at the higher (up to 1 mmol g^{-1} dry weight min^{-1} for glucose uptake in *Zymomonas mobilis* [15], or about 5 mmol g^{-1} dry weight min^{-1} for oxalate uptake in *Oxalobacter formigenes* [16]. Furthermore, kinetic analysis often helps one to find out whether more than one process is involved in the transport of a given molecule. This will be exemplified below with respect to excretion processes in Fig. 10 and Table 4 for a combination of a carrier-mediated and a diffusion-controlled transport.

2.3 Energetic Aspects of Carrier-Mediated Transport: The Concept of Coupling

The implementation of transport steps into concepts of metabolic engineering essentially requires evaluation of their demand in energy which must be provided by the cell as a driving force. Transport by diffusion (or facilitated diffusion) does not need an input of metabolic energy in order to take place, since it occurs with a negative change in free energy ΔG. This means that it is merely driven by the concentration difference of the transported molecules across the membrane. In biological systems, however, transport reactions frequently lead to the accumulation of a given molecule in the cell. As an endergonic reaction, this "uphill" transport is characterized by a positive ΔG, which means that it will never occur spontaneously. In order to take place, it must be "coupled" to another reaction with a sufficiently negative ΔG, so that the overall change in free energy of the coupled reactions is negative. Figure 5A elucidates the basic principle of coupling a "driving" (exergonic) to a "driven" (endergonic) reaction. The endergonic reaction on the right side will only take place if it is coupled to the exergonic reaction on the left. Compounds which disrupt this interrelation are therefore called "uncouplers", e.g. acetic acid as by-product of processes (see Sect. 3.3). The system will work from left to right only if the negative value of ΔG of the exergonic reaction is greater than the positive one of the endergonic reaction, otherwise it will work in the opposite direction. If the positive ΔG of the reaction on the right side increases (e.g. by accumulation of product D) to a value balancing that of the reaction on the left side, the exergonic reaction will be stoped by the coupling mechanism. A typical example for this kind of regulation is the fact that the activity of the ATP-synthase (= ATPase) depends on the actual phosphorylation potential of the ATP/ADP

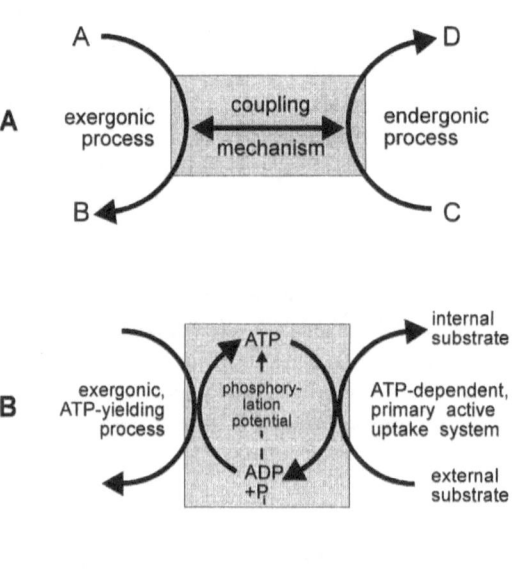

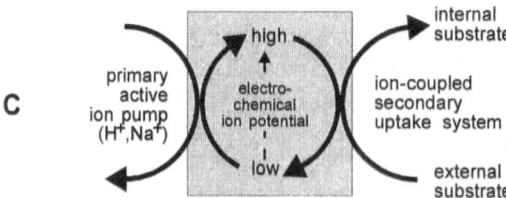

Fig. 5A–C. The principle of energy coupling in membrane transport: **A** the underlying concept is exemplified by the mechanism of energy coupling in; **B** primary and; **C** secondary transport. For the definition of transport mechanisms see Fig. 2 and text

system. In *E. coli*, for example, this primary system can work in both directions, i.e. synthesizing ATP under aerobic conditions if the respiratory chain provides a high electrochemical proton potential as a strong driving force, and in the direction of ATP hydrolysis under anaerobic conditions if ATP is provided by substrate level phosphorylation and the protonmotive force is comparably low (cf. also Fig. 7).

Many different biochemical and molecular concepts have been suggested to explain the principles of transport coupling in mechanistic terms. Currently, only two main mechanisms are known: chemical coupling (Figs. 5B and 6, primary transport) and coupling by ion currents (Figs. 5C and 7, secondary transport). Consequently, these two possibilities are connected with two types of general "currency" of energy: ATP, phosph*enol*pyruvate (PEP) and a few other high energy compounds on the one hand (chemical coupling), and electrochemical ion potentials on the other (coupling by ion currents) [4, 7, 11, 17–19]. Figures 6 and 7 give two examples for the two types of coupling. In Fig. 6, primary coupling connects the ATP-generating F_1F_0ATPase from the plasma membrane of a bacterial cell to the binding protein-dependent substrate uptake system which is driven by the free energy of ATP hydrolysis. In the same figure,

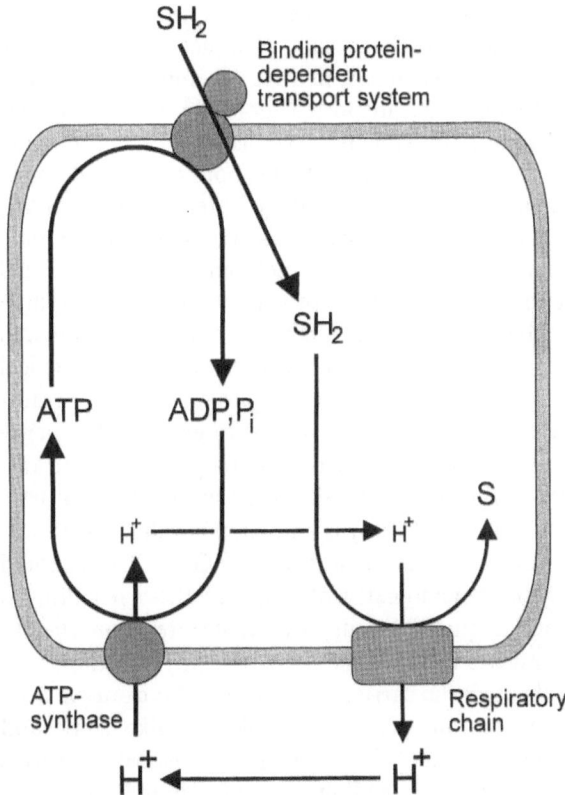

Fig. 6. Primary coupling of transport reactions. In a bacterial cell, the membrane-bound F_1F_0-ATPase synthesizes ATP at the expense of the electrochemical proton potential, which was created by the (primary) proton extrusion catalyzed by the respiratory chain. ATP is then used by a (primary) binding protein-dependent substrate uptake system which delivers reducible compounds to the respiratory chain. The ATPase and the binding protein-dependent system are linked via ATP (primary coupling)

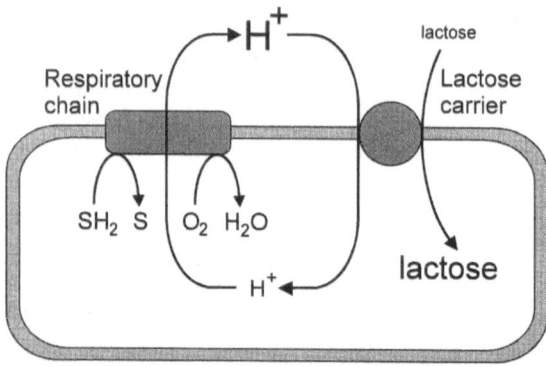

Fig. 7. Secondary coupling of transport reactions. The electrochemical proton potential in a bacterial cell, which is created by the action of the respiratory (cf Fig. 6), is used by the lactose uptake system (secondary symport) to accumulate lactose against its own concentration gradient. The respiratory chain and the lactose uptake system are linked via the electrochemical proton potential (secondary coupling)

the substrate is used by the respiratory chain to deliver redox energy for the active extrusion of protons, thus creating an electrochemical proton potential which, via secondary coupling, is the driving force for the membrane-embedded ATPase. Figure 7 exemplifies the typical situation for a secondary carrier, in this case the lactose uptake system of *E. coli*, which mediates accumulation of lactose in the cytoplasm driven by the electrochemical proton potential.

Provided the mechanism of a given transport reaction is known, its energy cost can be calculated. Transmembrane movement of molecule S across the membrane by a primary mechanism as shown in Figs 5B and 6 would need a certain number of ATP molecules per transported molecule or ion. Whereas for some ion-transporting ATPase the stoichiometry is in fact 1, for many primary systems stoichiometries different from unity, i.e. both above and below 1, have been determined. The energy cost of a secondary system (Figs. 5C and 7) depends both on the type of the coupling ion (e.g. protons, sodium, potassium, inorganic, anions, carboxylic acids), the coupling stoichiometry and the actual size of the gradients of those ions involved in the transport event. Since a large variety of mechanisms is possible, the transport mechanisms must be known in detail for a correct evaluation of its energy cost. Table 1 summarizes typical values for the energy cost of some representative examples of secondary transport reactions in terms of electrochemical driving forces. In general, the energy cost can be expressed or transformed into a certain number of protons/transported molecules. If the number of protons per ATP used by the ATPsynthase for the synthesis of ATP in the cells under study is known, this value can also be expressed in fractions of ATP or in an equivalent value of ΔG [4, 20]. It should be mentioned that the energetic stoichiometry of many of these reactions is actually not that simple, since frequently transport events are connected to different kinds of "losses". These are called "leaks", i.e. pathways for the transported molecules other than through the carrier proteins (see, for example Sect. 4.3, the transfer of partially membrane-permeable molecules like lactic and acetic acid, as well as isoleucine), and "slips", i.e. mechanistical (intrinsic) imperfections of the carrier systems leading to an altered transport stoichiometry and thus to energy losses [7]. Furthermore, a certain combination of transport reactions in a membrane may lead to effective waste of metabolic energy, a phenomenon which is called "futile cycling" (Sect. 4.3).

2.4 Regulation of Transport Activity by the Cell

Transport systems take part in a variety of functions which are essential for the cell, including uptake of nutrients, export of metabolic end products and signal processing. For an optimum function, these processes have to be tightly regulated. This, for example, includes adaptation to different environmental conditions, such as lack or presence of nutrients in widely varying concentrations, or the necessity of selecting between carbon sources of different metabolic quality. Furthermore, the activity of transport systems has to be adapted to the demand for maintaining homeostatic conditions for several ions in the cell, such as K^+, Na^+, H^+, Ca^{2+} and others, as well as for maintaining a certain steady-state in the energetic situation of the cell (phosphorylation potential or energy charge), also including the redox balance. There are at least four different levels of regulation of carrier activity (Table 3): (i) the transport activity simply depends on the availability and concentration of transport substrates and competing ligands in relation to their affinity for the carrier protein, (ii) it may also be influenced by parameters directly modulating kinetics of a carrier system such as intra- and extracellular pH, availability of cotransport ions, ion gradients and membrane potential, or by the medium osmolarity; (iii) a further level of control in the short-time range is regulation of the level of activity by interaction with particular effectors which may be internal metabolites (e.g. cyclic AMP), intracellular regulatory proteins or components of other transport systems (e.g. the HPr component of phosphotransferase systems [21]; (iv) as an adaptive mechanism to changes in the growth conditions, many carrier systems are effectively regulated on the level of expression (see below). The different levels and mechanisms of regulation are embedded into global regulatory networks

Table 3. Different levels of modulation/regulation of carrier activity

Level	Principle	Influence on	Time scale
1	availability of transport substrates, presence of competing ligands	carrier activity, apparent K_M	short (ms)
2	modulation of kinetic parameters and driving forces (pH, cotransport ions, ion (Na^+, K^+) gradients, ΔpH, $\Delta\Psi$)	carrier activity, apparent K_M	short (ms)
3	presence of effector ligands (allosteric activators and inhibitors)	V_{max} and/or K_M	short (ms-s)
4	expression of carrier proteins (induction and repression)	carrier concentration	long (min-h)

within the cell by which complex answers to particular conditions are coordinated, i.e. growth limitation by different substances, such as carbon sources, nitrogen sources, essential cations (e.g. potassium, iron, manganese) or anions (e.g. phosphate). Furthermore, crosstalk between transport systems and cytoplasmic enzymes can be very complex, according to the preference of a particular organism for particular substrates, thus providing a hierarchy of catabolic sequences. This means, for example, that in the presence of glucose and galactose, *E. coli* will first use glucose and then galactose ("glucose effect" or diauxic growth).

Many different mechanisms exist for catalysing uptake of molecules and ions as well as various types of coupling (Sects. 2.1 and 2.2). Frequently more than one type of uptake system for a certain molecule is found in a particular organism. In *E. coli*, for example, there are at least seven different uptake systems which accept D-glucose and D-galactose, or at least five uptake systems both for the anionic amino acids glutamate or aspartate and for aromatic amino acids. The diversity in mechanisms can be rationalized by differences in the availability of different transport substrates, as well as by different energetic demands of metabolism and transport. The observation that frequently a multitude of uptake systems for the same substrate is present in bacteria can be explained in physiological terms [22]. The main reason is the fact that they are used for different purposes. Only one (or often even no) uptake system for rare and unusual substrates may be found, or a very simple uptake system if the substrate is constantly available in high concentrations in the surroundings. In general, multiple uptake systems are present if the respective molecule is an important substrate for the cell in a given habitat, and if it is present in widely varying concentrations. As a rule, in case several systems for a given substrate are present, there is at least one constitutive system with comparably low affinity (high K_M) but high capacity (high V_{max}). Additional inducible systems, which are frequently induced by their major substrate, are provided for "cases of emergency" and in general possess high affinity (low K_M) but low capacity.

Primary transport systems normally have a high substrate affinity and are often practically unidirectional, thus allowing very high accumulation ratios. In contrast, secondary systems are in general reversible, and they may thus lead to leakage of the transport substrate in situations of low external substrate or low energy. The simultaneous presence of both types of system for a given molecule or ion may thus create energy-wasting futile cycles (see Sect. 4.3). Furthermore, secondary uptake systems are coupled to the membrane potential and/or ion gradients. Accumulation ratios (internal/external) achieved by secondary systems therefore, in general, do not exceed 10^2–10^3. The driving force of primary uptake systems, on the other hand, provides enough free energy for high accumulation ratios, which may be necessary if a given molecule or ion is present in the environment at very low concentration. ATP-driven uptake of maltose in *E. coli*, for example, causes up to 2×10^5-fold accumulation in the cell. Other aspects for the occurrence of several different uptake systems are specific regulatory demands under varying metabolic conditions. At least three different uptake systems for the amino acid proline, for example, are found in enteric bacteria. The reason for this situation is the fact that, besides a precursor in protein synthesis, proline is an important osmoprotectant, which is accumulated to high internal concentrations under hyperosmotic stress. Thus, the requirement of widely different activity (slow for anabolic purposes, extremely high for situations of osmotic stress), affinity and especially regulation characteristics requires the combination of different uptake systems.

3 Methodical Aspects

There are obviously very different levels of analysis of a given transport system. The simple disappearance from the medium, i.e. the metabolic use of a nutrient known to be principally membrane-impermeable, as well as the appearance of a metabolite with these properties as an end product can in general be safely interpreted on a purely phenomenological basis as being mediated by a membrane-bound carrier system. This may even lead to a consistent model of metabolism, although the transport steps per se are not at all considered. Numerous publications use this type of approach.

At a time when the knowledge of cellular metabolism on the molecular level is constantly increasing, a more detailed picture of the mechanism and significance of transport processes should be achieved. The transport process can be analysed and treated separately within a metabolic scheme, provided methods are available which allow the distinct measurement of the particular reaction of disappearance of a molecule in one compartment and appearance in the other, separated from interfering influences by the metabolism or by chemical reactions. This should ultimately lead to an appropriate consideration of the functional significance of a given transport reaction for a metabolic process with respect to fluxes and structured models.

An ambitious aim in biotechnological terms then is to use the knowledge of a given transport reaction for a purposeful modification of transport activity, efficiency and regulation. In order to reach this goal, on the one hand a detailed biochemical analysis of the mechanism of the transport system in kinetic and energetic terms is necessary. On the other hand, the molecular access to the genes coding for the corresponding transport systems is indispensable for a purposeful redirection and/or engineering of the cell metabolism, suitable for optimised production of a given metabolite.

Since the purpose of the present article is mainly the contribution of transport reactions to metabolic fluxes, the molecular biology aspect will not be treated in detail here (see [23]). The emphasis clearly lies on the appropriate tools for a quantitative analysis in kinetic and energetic terms, as well as the suitable approaches for a correct interpretation of these experimental results.

3.1 Experimental Analysis of Transport Processes

The quantitative analysis of transport reactions is based on the time-resolved determination of the concentration of the transported substances in one or both of the transport compartments (inside/outside, cytoplasm/medium, cis/trans). A variety of methods is available in order to analyse the dependence of transport on time (kinetics), energy input/output (energetics) and on particular structural and functional properties of the carrier protein (molecular mechanism). This can be done by quantitating the external and/or internal concentration of the transported molecule by chemical analysis, e.g. by HPLC and/or spectroscopy, by enzymatic analysis or by NMR. Furthermore, the transport reaction can be followed by using radiolabeled substrate molecules, the flux of which from one compartment into the other is monitored.

In several of these methods, the internal (cytoplasmic) concentration of the transported molecule has to be determined accurately. This, for example, holds for uptake systems which, due to a low external affinity (K_M, K_t) or low capacity (V_{max}), do not lead to a substantial decrease in the external substrate concentration within reasonable times in terms of kinetics, and thus do not give statistically significant experimental values. When studying export reactions, it is usually necessary to quantitate the internal concentration of the transport substrate, in order to analyse the excretion kinetics reliably. In general, the measurement of internal substrate concentrations needs separation of internal and external compartments, except when the compound is present solely in the cytoplasm, e.g. ATP, a situation, however, which is not relevant for studying transport.

The suitability of a particular method for this kind of analysis depends on the experimental constraints. These may be the time resolution necessary for separation of the compartments, or for stopping metabolic reactions, the stability/fragility of the cells used, the question of whether analysis of the external medium is also necessary, the quantity of the cells to be analysed etc. The simplest method used for separation of internal and external space is filtration of a cell suspension, which is often sufficiently fast (seconds) and quantitative. However, it is sometimes complicated to retain simultaneously the external medium for analysis, since the cells on the filter must be washed after the initial separation. If both cells and medium should be analysed, centrifugation methods are more appropriate. Simple centrifugation fulfils this requirement, although the time needed for separation of the two compartments is rather long (minutes) and the contamination by external medium in the cell pellet is substantial. A significant improvement in the basic method of separation by centrifugation is the method of silicone oil centrifugation [24–26]. The cell suspension is layered on top of a silicone oil layer, which in turn is underlaid by an inactivating solution, e.g. perchloric acid or detergent solution. The density of the three layers must be different and is chosen such that during (fast) centrifugation the cells are quickly sedimented (seconds) into the inactivating layer. Since the cells become stripped of the external medium during passage through the silicone oil, significant contamination from the external medium is avoided.

For particular purposes, e.g. for high time resolution, special procedures have been developed which are based on quenched-flow techniques. Cells and medium are mixed in a specially designed mixing chamber and are subsequently forced along a tube the dimensions of which determines the time of reaction. The reaction is stopped by separating the cells from the medium, e.g. by rapid filtration, or by instant cooling, e.g. in methanol at $-40\,°C$ (methanol spraying technique), and consecutive separation of the cells by centrifugation or filtration [27]. Fast sampling techniques are especially important when studying transport reactions in bioreactors (see Sect. 3.4)

For the typical time scale of transport reactions across cell membranes (seconds to a few minutes), NMR techniques are generally not well suited. Although internal and external molecules can, in principle, be separated by applying shift reagents [28], the time resolution of NMR spectroscopy is not sufficiently high for studying metabolite transport. Of course, like other analytical methods, NMR has been used to analyse metabolite concentrations after separating the internal and external space, which can also be applied to the study of metabolite transport [29]. By developing new techniques for the application to transport reactions, however, it has been shown that NMR spectroscopy can be used successfully for directly studying membrane transport processes [30], e.g. by saturation transfer NMR. These methods have recently been used to analyse sugar as well as ethanol transport in *Zymomonas mobilis* [31, 32].

There are various problems and pitfalls related to these kinds of methodical approaches. (i) As will be discussed below (Sect. 3.2.), it is often difficult to separate unequivocally a transport process from a connected metabolic reaction, at least when using intact cells in the experiments. (ii) If a given transport reaction under study is relatively slow or does not lead to significant accumulation in one of the compartments, it is frequently difficult to discriminate transport from unspecific binding. This, of course, depends significantly on the type of substrate, amphiphilic and hydrophobic compounds being especially prone to unspecific binding to membranes. Various methods have been applied to discriminate the contribution of unspecific binding, e.g. inactivation of the cells or washing with high concentrations of (unlabeled) substrate at low temperature [33]. (iii) A well-known difficulty in this kind of study is the problem of the correct determination of the internal (cytoplasmic) volume which is necessary for the calculation of internal concentrations. Although sophisticated methods have been worked out to quantitate the internal and external spaces in cell pellets and especially in connection with the method of silicone oil centrifugation (see above), there are several reasons for misinterpretation. Certain compounds used to discriminate internal and external space based on their basic impermeability may, in addition, be absorbed to the cell wall. Furthermore, the increasing number of (unspecific) drug extrusion systems discovered recently in prokaryotic and eukaryotic cells makes these methods questionable to some extent, since they are also known to transport the compounds used in these analyses of cell volume [34–36].

Another essential aspect when elucidating the mechanisms of a given transport system is the analysis of its energy requirement. Besides simple facilitated diffusion, transport processes need the input of metabolic energy (see Sect. 2). Although the correct quantitation of the kind and amount of energy necessary to drive a particular transport event is of utmost importance for integrating the transport step into a structured model of metabolic fluxes, this question frequently needs a relatively complicated biochemical analysis to be solved. For the secondary carrier proteins, for example, the kind of transport (uniport, symport, antiport), the kind of coupling ion (H^+, Na^+), and especially the coupling stoichiometry must be analysed [6, 7]. Some examples of carrier systems analysed by this kind of approach have already been described in Table 1.

It has been mentioned above that frequently more than one transport system is available in a cell for a given molecule (see Sect. 2.4). This significantly complicates any kinetic and energetic analysis and in general makes the application of molecular techniques necessary to separate unequivocally the different transport systems by knocking out one or the other by mutation. Molecular techniques are also extremely helpful for identifying whether a particular reaction exerts a significant share of flux control on the overall metabolic process (see Sect. 3.5 and [23]).

3.2 Discrimination Between Diffusion and Carrier-Mediated Transport

Frequently, when starting to investigate a particular transport process, it is a fundamental question whether the transmembrane movement of a given molecule is mediated mainly by a diffusion-controlled or by a carrier-mediated process. This question is in fact essential since it is obvious that diffusion cannot be influenced purposefully by changing the process or engineering the organisms, except by modification of the concentration of the molecule of interest at one side or the other of the membrane. Furthermore, important consequences for the process have to be considered with respect to energy requirement (zero in the case of diffusion), with respect to a possible kinetic limitation and with respect to the expected product concentrations at the cis or the trans side (the side of generation of the product or addition of the nutrient).

In particular, excretion has frequently been erroneously interpreted as being caused by diffusion, by "membrane leakage" or the involvement of "unspecific membrane pores". In many cases, this has proven not to be true [1, 12]. Plasma membranes of living cells are in general extremely tight for most molecules, as can be seen from the fact that a significant electrical potential across these membranes is kept more or less constant which would collapse by a significant unspecific ion movement. Purely passive processes are generally responsible for uptake or excretion of a given molecule only if it is significantly hydrophobic, or if the cells are in fact severely damaged. In spite of this statement, however, care must be taken in every single case to assign a particular mechanism to an observed transport process. As an example, it has recently been shown for alanine excretion in recombinant *Zymomonas mobilis* strains that the excretion, surprisingly, was a purely passive process [37].

Based on the methods discussed in Sect. 3.1, several tests can be applied to differentiate experimentally a diffusion-controlled process from carrier-mediated movement of a molecule. The basic properties of a catalysed process, i.e. the phenomena of saturation, substrate specificity and inhibition by particular reagents, are generally appropriate tools to decide this question. This basic approach, however, is frequently helpful only on a theoretical basis due to several reasons, (i) these tests cannot easily be applied to excretion processes, since it is in general not easy to manipulate purposefully the internal metabolite concentration, (ii) under certain circumstances, a relatively fast diffusion processs is difficult to discriminate from a carrier-mediated transport reaction when the latter is characterized by a very low affinity, (iii) a further objection to this basic kind of discrimination is the fact that, in general, in the case of diffusion-controlled processes, enzymatic reactions of the consecutive metabolism may be directly connected to the observed transport reaction. This set-up would, erroneously, make a diffusion step look like a carrier-catalyzed reaction. Unequivocal tests for the presence of carrier-mediated reactions are particular experimental observations of trans-acceleration or a so-called countertransport maximum [3, 38] (see below), which, however, generally need the use of labeled transport substrates and are again difficult to apply to excretion processes. A decisive discrimination in favour of a carrier being responsible for the transport reaction is, of course, the availability of a transport-negative mutant.

Thus, whereas the question to discriminate diffusion and carrier-mediated transport can generally be solved for uptake reactions by basic kinetic and/or molecular approaches, this decision is much more difficult to obtain in the case of metabolite excretion. Again, a solution would be the availability of an excretion-negative mutant strain which, for example, has been achieved in the case of lysine excretion in *C. glutamicum* [39]. But the methods to screen for such kinds of mutants are generally difficult and time-consuming. A useful biochemical approach then would be to measure carefully internal and external concentrations of the given substance and to correlate transmembrane flux and concentration gradient [40, 41]. In general, based on a sufficiently large data set of this kind of measurement, it is possible to discriminate a saturable process, which is frequently regulated by the cell, from simple passive diffusion.

3.3 Kinetic Analysis of Transport Processes: Theory and Application

As discussed above, for kinetic analysis of carrier reactions the basic formalism derived for kinetic analysis of enzyme reactions is adequate for use (see Sect. 2.2). In general, transport reactions are characterized by measure of their substrate affinity (K_t or K_M) as well as by their transport capacity (V_{max}). The question now dealt with in this section is the availability of kinetic tools for analysing transport reactions to the extent that they can be assigned to the appropriate classes and mechanisms (see Sect. 2).

The tools of enzyme kinetics, as discussed in numerous books and reviews [e.g. 42–44], have been successfully translated into carrier kinetics [3, 45–49]. In addition to basic kinetics, specific properties of carrier proteins embedded in the surrounding membrane have to be considered, such as the dependence on the physical state of the membrane, the dependence on transmembrane osmotic gradients, the influence of local charges from the membrane surfaces and, most importantly, the influence of vectorial instead of only scalar parameters, such as gradients of ions and substrates, and the electrical gradient. Although thereby adding a significant number of additional parameters to be considered in a kinetic model including transport reactions, they can frequently be treated by relatively simple formalisms, at least formally.

If diffusion is involved, the corresponding thermodynamic (electrochemical potential) and kinetic (Fick's first law of diffusion) formalisms have to be considered in order to describe properly diffusion-controlled fluxes across membranes [3]. Kinetic treatment of membrane channels (pores) with respect to maximum conductance and half-saturation constant [3, 45, 50] has been elaborated in detail, mainly because of their importance in signal transduction. Channels are not very significant for the topic discussed here, i.e. for membrane transport of metabolites, at least in kinetic terms. An exception are the pore proteins of the outer membrane of bacteria, the so-called porins, which, according to present knowledge, are generally not rate-limiting in uptake or excretion of metabolites [51].

The simplest mechanism of carrier-mediated transport is secondary uniport or facilitated diffusion, which can be treated by the formalisms of enzyme kinetics as described above (see Sect. 2.2) including competition by alternative subtrates and inhibition phenomena. It should be pointed out, however, that the derived kinetic parameters depend critically on the particular experimental conditions, for example whether a zero-trans experiment (i.e. zero substrate concentration at the trans side) or a study of equilibrium exchange (i.e. identical concentration of transport substrate at both sides) is carried out. Consequently, unequivocal proofs for transport being carrier-mediated are based on exactly this kind of observation, i.e. the influence of a substrate at the trans side on kinetic parameters at the cis side. This means that the transport rate is in general stimulated by the presence of saturating substrate concentrations at the trans side in comparison to a zero-trans experiment [3, 38]. One of the best-studied systems of this type is the glucose carrier (facilitator) in higher eukaryotes [52–55].

The simple conditions of carrier kinetics may become more complex if additional properties have to be considered, either of the carrier protein, e.g. an asymmetric kinetic behaviour with respect to the two membrane sides, or of the substrate, e.g. the involvement of electric forces if the substrate is charged. The mechanistic of kinetic explanation of carrier mechanisms becomes significantly more complicated in the case of coupled transport systems, i.e. secondary and primary active transporters, which are able to move the substrate against its electrochemical gradient (see Sect. 2.1). Whereas the thermodynamic equilibrium of this kind of transport system is governed by the sum of all forces involved and by the coupling stoichiometry of these forces (see Table 1), it is still useful to treat formally the kinetic behaviour by the same simple approaches as mentioned above. In general the multitude of external and internal kinetic constants and other values necessary for a complete kinetic description are not readily available. It must be critically taken into account, however, that, based on the principle of coupling, the kinetic properties of the flux under study may change when the situation of the coupled force, e.g. that of the cotransported ion, is changed (see below).

An important question now concerns the relevance of the kinetic description of a carrier system for considerations concerning metabolic flux and for a biotechnological application. In fact, besides the basic characterisation of the mechanism of a particular carrier system in kinetic and energetic terms, it is essential to elucidate quantitatively whether and to what extent kinetic or energetic parameters which may be influenced by conditions of the bioprocedure will significantly modulate a given transport reaction. Changing the medium pH, for example, may have a complicated spectrum of different consequences. (i) The actual pH at the external membrane surface influences the activity of membrane bound enzymes, e.g. carrier proteins, by altering the ionization state of particular amino acid residues or by altering the conformation of a protein. This reflects to the "true" pH optimum of an enzyme or carrier protein (ii) A significant change in the external pH will normally also change the internal pH value to some extent. It is known for a variety of transport systems, especially those for amino acids, that even a moderate decrease in internal pH leads to full inhibition of these systems [56, 57]. (iii) Protons are coupling ions in many transport processes (H^+-coupled secondary systems, Sect. 2.1), and in this case they will directly modulate the transport energetic and kinetics. (iv) If the transport substrate is protonizable, generally only one form of the substrate is accepted by the carrier protein, the actual concentration of which is thus governed by the medium pH. Well known examples are phosphate, sugar phosphates, carboxylic acids, e.g. succinate or citrate, and amino acids, e.g. glutamine. (v) A change in the pH gradient across the cytoplasmic membrane influences the actual electrical gradient (membrane potential), since cells try to keep the protonmotive force (the electrochemical proton potential, i.e. the sum of the pH gradient and the membrane potential constant [58, 59]. The membrane potential, on the other hand, is the essential driving force in many transport processes (see Sect. 2.1). This example may illustrate the complex consequences, as measured by kinetic techniques, on carrier function which may arise and which have to be considered when altering only one single parameter in the medium.

A variety of typical events occurring during the procedures may further illustrate this statement. If the concentration of a membrane-permeable weak acid, e.g. acetate which can act as an uncoupler [29, 60, 61], changes during growth, it significantly reduces the driving force of electrogenic transport systems, e.g. uptake of substrates such as carboxylic acids or carbohydrates. Besides the direct effects of changing the external pH as explained above, even titration of medium pH changes, e.g. by addition of NaOH, may influence the activity of transport systems if the substrate flux is coupled to that of Na^+ ions or if it is sensitive to changes in osmolarity. Similar considerations are true for primary active systems, which may depend critically on variations of the cytoplasmic concentration of ATP or of phospho*enol*pyruvate.

Another important point, which is not even considered here, is the fact that the activity of many transport systems is not only modulated by functional parameters as discussed above, but is also regulated by the cell both on the level of activity and on the expression level. Consequently, the most useful initial procedure for kinetic purposes is to model the transport reaction as simply as possible. Then, in order to evaluate which changes in the metabolic or the process conditions may significantly influence the activity of the transport system under study, as much additional information as available on cotransport ions, antiport substrates, pH-dependence, electric behaviour, and energy coupling should be taken into consideration. Finally, more and more information on the regulation of both levels, i.e. activity and expression, may be considered in order to evaluate the significance of the model developed for different metabolic situations. As a consequence of this complex situation, it has to be expected that the dependence of transport activity on the concentration of the transported molecule may sometimes be not that simple, i.e. it may include threshold values due to regulation (an example is shown in Fig. 10) or complicated kinetic behaviour due to a multiplicity of systems.

Several examples illustrating this situation in more detail are described throughout the text. The multiplicity of sugar transport systems in yeast and their different energy dependences are discussed (Sect. 5), as well as considerations, whether a facilitator transport system (e.g. glucose uptake in *Z. mobilis* and in *S. cerevisiae*) or an energy-dependent accumulative system (e.g. PTS in *E. coli*) offers advantages for particular situations during fermentation or for productivity (Sects. 4.1 and 5). Another example is provided by the description of the complex kinetics of excretion systems for various amino acids, such as lysine (Sects. 4.2 and 5) and isoleucine (Sect. 4.3), and their importance for the production of amino acids.

3.4 Analysis of Transport in Bioprocesses

From a microbiologist's point of view, there are two principle approaches to analysing membrane transport processes – by applying basic kinetic methods to stationary cells in a test tube, e.g. the time-resolved uptake of radiolabeled substrates as described above, or by continuous measuring of substrate and product concentration in reactors. The two approaches can of course also be combined which may lead to qualitatively new interpretations (see example in

Sect. 5). Analyses by using typical biochemical (kinetic) methods have already been described in Sect. 3.1 and 3.3. On the other hand, it is often interesting to derive the activity of a transport system directly on line [62]. The two different methodical approaches have their advantages and disadvantages. It is obviously easier to elucidate a particular transport mechanism by experiments in the best tube, since all external and some internal parameters of interest can be changed at will. But for the true physiological situation with respect to regulation on level of activity and expression it may be more relevant to study transport in a reactor under defined conditions of growth and metabolism. Results obtained from transport analysis with stationary cells defining a particular biochemical mechanism (e.g. uniport, symport or antiport) can be directly used to explain the carrier's function in a reactor, i.e. the substrate affinity, the type of energy coupling, or the transport stoichiometry. The same does not hold for the measured transport activity. Because of different conditions with respect to both the internal (cytoplasm) and the external (medium) compartment, and because of a different metabolic state, e.g. exponential or limited growth, the measured values for actual or maximum transport rate of a given carrier system may vary significantly between the reactor and the test tube. An important consequence of this situation is the fact that kinetic data with respect to specific transport activities should not simply be taken from the literature but should be determined in the actual process setup.

For a quantitative study of transport reactions (uptake of nutrients and excretion of products) under defined conditions of growth it is necessary to control the external conditions, e.g. medium pH, dissolved oxygen, presence of trace elements, or medium osmolarity (see also [23, 63]). These and other parameters may have significant short- and long-term effects on the activity (e.g. pH, ion composition and osmolarity) and expression of carrier systems (e.g. oxygen saturation as well as concentration of nutrients and trace elements). Consequently, shake flasks are not well suited for this purpose. Batch mode under control of pH and oxygen provides a much better condition, although this is only true as long as the concentrations of all nutrients are in excess, i.e. above their apparent K_M (or K_S) values. Otherwise the situation becomes very complicated due to shifts in the growth rate, in the extent by which growth is coupled to transport reactions, by variation of expression of carrier systems, and possibly by the occurrence of conditions of overflow metabolism (see Sect. 4.2). In addition, formation and accumulation of by-products during batch fermentation, e.g. lactate or acetate, may influence the activity of transport systems (see Sects. 2.3 and 3.3.). The particular complication of a limitation by nutrients can be overcome by using fed-batch mode, which, however, is no solution to most of the other problems mentioned above, e.g. changing growth and medium conditions and accumulation of by-products.

The best-suited setup for studying transport processes in a reactor is a continuously operated stirred tank reactor at a steady-state (see [63]). In a steady-state, the uptake rate v_s of a given nutrient can be calculated directly from the dilution rate D, the concentration of biomass X, the difference in the concentration of the particular nutrient s in the reservoir C_{si} and its residual concentration C_s in the effluent:

$$v_s = D(C_{si} - C_s)/X.$$

The same holds for the rate of excretion of a particular product p which is related to the concentration of this product C_p in the effluent:

$$v_p = D C_p/X$$

The measured uptake rate v_s equals the V_{max} value of the uptake system only if its concentration in the reactor is far above the respective K_M value of this particular substrate. Otherwise the maximum uptake the $V_{s,max}$ can be calculated by

$$v_s = V_{s,max} C_s/(K_M + C_s)$$

provided the transport reaction follows Michaelis-Menten kinetics which is true in many cases. The same calculation cannot easily be used for product excretion, since in this case the actual internal concentration of the substrate must be known, as well as its internal K_M value, which is normally not available (see, however, in the case of lysine excretion, Sects. 4.2 and 5). By changing the process parameters (e.g. pH, temperature, concentration of ions and supplements, as well as the dilution rate and the connected values of the concentration of biomass and of the limiting substrate) the functional characteristics of a transport process can be analysed. It is obvious that this type of approach for elucidation of a transport mechanism is much more difficult and time-consuming compared to direct tests using cells in a test tube. Thus the study of the properties of transport processes during continuous cultivation is used mainly to test whether particular aspects of interest of a previously analysed transport mechanism hold true under the defined physiological conditions of growth or metabolism.

An interesting combination of the two methodical approaches (see above) is the use of a bioreactor for obtaining defined metabolic conditions connected with fast sampling techniques, as described in detail in [63]. Only those methods, however, which are able to discriminate between the compounds in the medium and those in the cytoplasm (or even in intracellular compartments of eukaryotic cells, such as mitochondria) are applicable to this purpose. This fact more or less restricts the choice to methanol-spraying techniques (see Sect. 3.1 and [63]).

The most serious methodical problem for this type of analysis is the fact that it is not easy to discriminate between transport and metabolism being the actual parameter to be measured. A prominent example is glucose metabolism in yeast where the kinetic separation of glucose uptake from consecutive phosphorylation has proven to be extremely difficult and thus prone to possible misinterpretations [33, 64–66]. Although this problem holds for all kinds of measurements with intact cells (see Sect. 3.3), it is most prominent in bioprocess studies since some methodical tricks as used in experiments with stationary cells, e.g. the addition of nonmetabolizable substrates, are not applicable. Thus a value for the transport activity obtained in this kind of measurement may not reflect in the maximum capacity of the transport reaction if a closely connected metabolic reaction is kinetically limiting, nor may it tell whether transport is limiting under the conditions studied. The only way around this problem is the use of independent qualitative information, whether nutrient uptake or product excretion is in fact the limiting step. Several experimental approaches in this direction are described in the next section (Sect. 3.5).

3.5 Transport Reactions as Putative Controlling Steps

Nowadays, the extent to which a certain reaction is a limiting step is frequently treated in terms of the metabolic control theory [67–72]. This theory is discussed in more detail in [23]. With respect to transport reactions, it may be understood in treating a given part of the metabolic network as being in a steady-state. The flux control coefficient of a particular reaction, in this case the transport step, is the quantitative measure of the extent to which a particular step (transport reaction) in the overall metabolism takes part in controlling the total flux. Similar considerations can be applied to control by metabolite concentrations.

Whereas sophisticated methods have to be applied to evaluate the true control coefficient of a given transport reaction [69, 73–75], in the context of this article it is of major interest to describe approaches in order to evaluate qualitatively whether a transport reaction in fact has a relevant (limiting) influence on the metabolism, at least under particular metabolic conditions. Provided that it has been shown that the flux of a particular molecule across the membrane is mediated by the action of a carrier system, there is still a broad gap between this basic finding and the knowledge of whether or not its capacity is limiting. This situation is analogous to enzymes and metabolic networks, where it has been found in many cases that detailed measurements of enzyme activity in cell extracts provides only a rough idea of the enzyme's significance in the metabolic web under a particular metabolic situation. The same holds for carrier systems. Even if the concentration of the substrate of a particular transport system on both sides (cis and trans side) is known, as well as its V_{max} and K_t (K_M) values, its actual activity may be critically influenced by regulation on the level of activity, e.g. by metabolic activators and/or inhibitors which are lost, diluted or inactivated in the extract.

There are several experimental observations which indicate whether or not a certain transport step is limiting in an overall metabolic sequence. This may be analysed in the case of nutrient uptake as well as product excretion by detailed measurement of the internal concentration in relation to the external concentration of the particular substance of interest. The observation of a significant internal concentration of the molecule taken up indicates that the uptake reaction is not the limiting step under these conditions, whereas the opposite holds true for carrier-mediated product excretion, i.e. a low internal value indicates that the excretion step is sufficiently fast. For uptake reactions, the basic approach used in the metabolic control theory to evaluate the control strength of a particular reaction can be applied, i.e. titrating the uptake reaction by addition of a specific inhibitor or a competing ligand which is not a transport substrate. This has been done successfully with mitochondrial membranes, where the control coefficients of enzyme complexes of the respiratory chain, of cytosolic and inner-mitochondrial enzymes, as well as carrier systems of the inner mitochondrial membrane have been quantified [73–75]. Another possible approach to overcome these problems is to analyse directly the transport capacity of the cells in the reactor by repeated *off-line* biochemical (kinetic) measurement of cells under conditions which guarantee a comparable metabolic and regulatory situation. An example of this methodical approach is described in Sect. 5. An attractive, fundamental approach would of course be the purposeful and controlled modulation of carrier expression by recombinant DNA techniques. This, however, not only needs the availability of the gene(s) of the carrier system in the organism under study, as well as appropriate molecular methods for their expression, but also ways to control the expression in a narrow range (a few percent) of activity above and below its physiological value. In general, only significant overexpression or full disruption of a particular gene can easily be achieved, which is often not a well-suited strategy. The consequences of the application of this kind of method, whether using biochemical or molecular approaches, i.e. flux modulation by alterations in the activity of a particular enzyme or carrier, are described in more detail in [23].

4 Significance of Transport for Biotechnological Processes

In this section, some examples will be discussed which illustrate quantitatively to what extent transport reactions may be important in biotechnological processes. This refers to the aspects whether the presence or activity of carrier systems may direct the use of particular substrates and the occurence of particular products, and in particular to which extent the yield and/or the production rate are influenced, due to effects on the kinetics and/or energetics of the overall process.

4.1 The Relevance of Nutrient Uptake
in Biotechnological Processes

Since most nutrients cannot passively cross the plasma membrane of cells, the presence of an appropriate uptake system is a trivial prerequisite for any bioconversion [76]. Provided a suitable uptake route is present, the kinetic and energetic properties of this particular transport system(s) may be important for the biotechnological application and especially for modeling purposes. Not only in the most simple case, i.e. limitation of growth or of metabolite production by the maximum capacity of the uptake system for carbon or energy sources, the significance of a quantitative knowledge of uptake systems becomes obvious. There are at least three different aspects of a particular kind of transport mechanism present which may critically influence modeling purposes or the further biotechnological application. (i) The functional properties of the transport mechanism may be important for designing an appropriate condition of growth and production and for the application of the correct parameters in structured models. This, for example, refers to the demand in cotransport ions, e.g. sodium in many systems, to the influence of external pH on uptake systems, to the presence of membrane-acting compounds during fermentation, e.g. acetate or ethanol, which may influence the activity of transport systems (see Sects. 2.3 and 3.3). Finally, uptake systems may have a specific functional construction by which they need particular countersubstrates at the interior of the cell for their proper function in uptake (e.g. in precursor/product antiport systems in bacteria [77]. (ii) In general, uptake of nutrients from the surrounding medium into the cell needs metabolic energy. It depends on the construction of an uptake system which type of energy is used (e.g. ATP, phospho*enol*pyruvate, chemiosmotic energy of ion gradients), which in turn poses specific constraints on the structure of the model to be considered since these energy sources must be provided in a balanced way by the central metabolism. It thus makes a significant difference in model building, whether a transport system uses phospho*enol*-pyruvate in stoichiometric quantities, like glucose uptake in many bacteria by the PTS-systems, or the electrochemical proton gradient which is provided with high efficiency by the respiratory chain in most aerobic bacteria, or whether it needs no metabolic energy at all, e.g. for glucose uptake by the glucose facilitator in bakers yeast. (iii) There are only few examples of nutrient uptake systems which are not regulated by the cell and which thus provide more or less unlimited access to the metabolic reactions in the cell, a well-known example being glucose uptake by the sugar facilitator in *Zymomonas mobilis* [78–80]. In most cases the activity of the energy-dependent and accumulative nutrient uptake systems is regulated by the cell, thus adapting the internal nutrient availability to physiological demands. These in turn do not automatically match the demands of a particular production process using the cell's metabolism. A close connection is often observed between important uptake systems and the regulatory networks of the cell, e.g. that of the response to nutrient limitation or the networks connected to energy and redox balance. An essential prerequisite, for example, in the concept of metabolic overflow which may be important in a number of conditions leading to product excretion, is a relatively weak regulation of uptake systems for carbon and energy sources (see Sect. 4.2). This was shown both in bacterial [81, 82] and also in yeast cells [82a]. In the latter case, the overflow situation was quantitatively described by a dynamic model [82b].

4.2 The Relevance of Product Excretion
in Biotechnological Processes

In view of the long list of low-molecular-weight substances produced by bacterial and eukaryotic cells, e.g. carboxylic acids, amino acids, antibiotics and vitamins, and in view of the fact that most of these products are not readily membrane-permeable, the relevance of carrier-mediated product excretion for biotechnological processes becomes obvious. Apparently, we are dealing with at least two distinguishable types of system, both from the physiological point of view and in the strategies to we used for optimization in biotechnological processes. On the one hand, we have products which use excretion mechanisms to which some physiological meaning can be assigned, e.g. excretion of organic acids (e.g. lactic acid) and alcohols, or organic acids (e.g. citric acid, oxoglutaric acid) as well as amino acids (e.g. glutamic acid) under conditions of overflow mechanism [1, 29, 81–84]. On the other hand, cells in which the metabolism is fundamentally redirected to products which are "normally" not excreted, e.g. some vitamins or essential amino acids, the question arises about the meaning of export mechanisms for these compounds. Either (i) these substances may cross the membrane by passive mechanisms which are not carrier-mediated because of extremely high internal accumulation due to the redirection of the cellular metabolism, or (ii) they are exported by routes which are normally not used for this purpose, or (iii) their excretion is catalyzed by specific excretion carriers, whose physiological meaning is so far unknown [1, 12]. In view of these considerations, the relevance of the discrimination between passive and carrier-mediated processes, as discussed in detail above (Sect. 3.2) becomes obvious again. There are important consequences for purposeful metabolic engineering, whether the export of a certain compound is a passive process or whether it needs a carrier protein to be mediated.

Furthermore, the latter examples, i.e. excretion by a carrier-mediated mechanism of unknown significance, may also have consequences for the properties of this system in terms of regulation. It is interesting as to whether (i) the carrier system normally transports other (physiological) substrates and is simply "misused" by the product under study, or whether (ii) the excreted product possibly uses an uptake system present in the organism for this particular substance. Examples of both cases are known, i.e. extrusion of toxic compounds from eukaryotic and prokaryotic cells [34, 35, 85, 86] or extrusion of amino acids in lactic acid bacteria growing on peptides [87]. In any case, the nature and the properties of these systems are of utmost importance in order to include them properly into structured models of the overall process.

The first specific excretion systems to be studied experimentally were those for lactate in a variety of bacteria [88, 89]. The particular significance of carrier-mediated lactate excretion by E. coli for the energetics of the bacterial cell has recently been studied in detail [29]. Besides efflux of several organic and amino acids in the process of precursor/product antiport systems which are restricted to particular types of metabolism [77], the excretion of amino acids by the well-known amino acid producing bacterium *Corynebacterium glutamicum* has been studied in more detail in recent years [1, 12, 90]. It has so far been shown for glutamate, lysine, isoleucine and threonine that their excretion depends on specific, energy-dependent and regulated excretion carrier systems. Generally speaking, the same trivial point as mentioned in the case of nutrient uptake (see above) may also be relevant to product excretion, namely the fact that simple lack of an appropriate export system may govern the applicability of a certain process in a particular organism. It is interesting to note that some organisms are preferentially used for production of certain compounds, e.g. coryneform bacteria for amino acids. It is not clear so far whether the main reasons for this are particular regulation patterns of anabolic pathways (for details see [23]), which may be suggested by the fact that *E. coli* is also extensively used for amino acid production after appropriate redirection of the intermediary metabolism, or whether this is in fact due to a certain preference in the ability to excrete these metabolites [1].

As an example, the biochemical analysis of lysine excretion in *C. glutamicum* led to a kinetic model (Fig. 8) in which the (cationic) amino acid is exported together with OH^- ions, which is equivalent to antiport of protons [40, 91]. Consequently, the activity of the lysine export carrier is modulated by a set of physiological parameters, i.e. the membrane potential (electrical potential), the pH-gradient and the lysine gradient across the membrane. Based on measurements of the relevant external and internal kinetic constants, a kinetic model describing quantitatively the dependence of the transport rate on these parameters has been developed [91]. Knowledge of these kinds of parameter provides a basis for analysis of the contribution of the excretion step to the overall flux leading to lysine production in *C. glutamicum* (for a detailed treatment of the

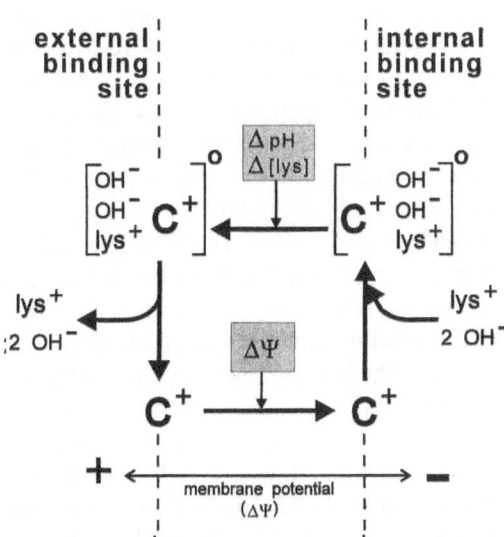

Fig. 8. Kinetic model of the lysine excretion carrier system as elucidated in *C. glutamicum* [91]. The substrate binding site of the carrier is supposed to be positively charged (C^+) and to reorient between the external and the internal faces of the membrane during transport. Excretion of lysine (lys^+) occurs in symport with two OH^--ions, and thus the net charge of the actual transport complex changes and the kinetic cycle of the carrier protein may be modulated at various partial reactions by the membrane potential ($\Delta\Psi$, positive outside), the pH gradient (ΔpH) and the lysine gradient (Δlys) across the membrane

basic metabolic situation [23]). Functional analysis of lysine flux under production conditions revealed that excretion may in fact be one of the steps of major flux control in lysine production by *C. glutamicum* [92, 93]. This was corroborated by the analysis of a *C. glutamicum* mutant which lacks a functionally active lysine excretion system [39]. Interestingly, a difference in the kinetic properties and in the regulation of the lysine excretion system between wild type strains of *C. glutamicum* and production mutants was found [94, 95]. In summary, it becomes clear that the combination of physiological, biochemical and molecular data on an export carrier system may have significant consequences both for the proper implementation of this transport step into structured models as well as for strategies to improve further lysine production by metabolic engineering [23].

It should also be pointed out that for excretion systems the integration into cellular regulation networks is very important for a proper understanding of their physiological function and activity, as has been discussed already in connection with uptake systems. As an example, overflow metabolism is actually a description of a certain kind of regulatory status of the cell. It has been shown, at least for amino acid excretion, that the energy state and the availability of carbon may have a direct regulatory influence on the activity of excretion systems [1, 41, 83, 95]. Another example is the, so far, poorly understood connection between limitation of growth and/or metabolism and amino acid overproduction [1, 84, 96–99] (see also Sect. 2.4 and 4.2).

4.3 Interfering Combinations of Fluxes

In view of the multiplicity of possible carrier-mediated as well as passive fluxes of ions and metabolites, situations are possible which lead to an unproductive combination of different transmembrane movements of molecules and ions, thus creating so-called "futile-cycles". These cycles may be futile in two respects, i.e. waste of energy and waste of carbon source (or product). The original meaning refers to a combination which simply leads to dissipation of energy and thus to a conversion of metabolic energy (redox energy, ATP, electrochemical potentials) into heat. Classic examples are the combination of a primary pump, e.g. redox-driven proton extrusion by the respiratory chain, and a leak, e.g. by the presence of an uncoupling substance such as acetate (Fig. 9) [29, 60], as well as combinations of ion transporting systems in opposing directions, e.g. K^+ uptake by the K^+-ATPase and concomitant K^+-efflux via a secondary system [100, 101]. The possible presence of this kind of combination of counteracting fluxes is a basic problem in cell energetics which is undoubtedly also of fundamental interest in approaches to optimize the efficiency of producing strains. A strong indication for the fundamental significance of this fact is the discrepancy between the theoretical and the actual value of Y_{ATP} [20, 102, 103]. This discrepancy is not yet understood and is most likely due to exactly those type of futile cycles involving different transport systems [84, 96, 100–107]:

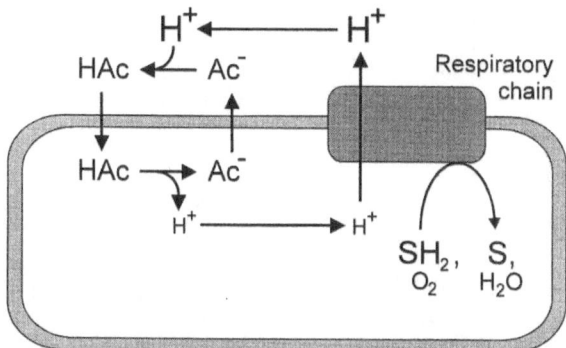

Fig. 9. Futile cycling mechanism exemplified by the combination of primary proton extrusion (respiratory chain, see Figs. 6 and 7) and passive proton movement across the plasma membrane mediated by acetic acid. Acetic acid, which is membrane permeable both in its protonized and its unprotonized form (although to different rates), leads to backflow of protons through the membrane. Thereby the chemical energy of the reduced substrate SH_2 is converted into heat

Besides the general importance of futile cycles, more specific examples are likely to occur in connection with uptake and excretion transport processes during metabolite production as discussed in this article. Overflow metabolism, for example, occurs under situations of energy surplus, and thus economic balancing of the energy budget is not necessary under these conditions. On the contrary, energy waste by operating futile cycles may be essential for optimum productivity under certain circumstances, as has, for example, been suggested for amino acid (lysine) production in *C. glutamicum* [96, 99, 108]. Thus, even a controlled introduction of futile cycles may be necessary in a strategy of purposeful metabolic design.

The metabolic situation as described here, on the other hand, indicates that product excretion under these conditions is likely to occur irrespective of the possibility of counteracting fluxes. A typical example connected with amino acid production is the excretion of isoleucine in *C. glutamicum* (for details of the metabolic situation, see [23]). This organism possesses an energy-dependent uptake system of relatively low activity. Isoleucine also crosses the membrane by passive diffusion due to its hydrophobic character. In addition, *C. glutamicum* can excrete isoleucine by an energy-dependent, tightly regulated excretion system [41]. The combination of these partially counteracting fluxes is shown in Fig. 10 for varying internal isoleucine concentrations in the presence of low external isoleucine. The activity of the isoleucine uptake and the excretion system is regulated by the internal isoleucine concentration both on the level of activity, thus showing significant activity only above a certain threshold concentration (Fig. 10), and on the level of expression. Consequently, the concomitant activity of the three transmembrane fluxes leads to a relatively complicated pattern (Table 4). It becomes obvious that at high external isoleucine concentrations (lower part of Table 4), i.e. in later stages of isoleucine bioproduction

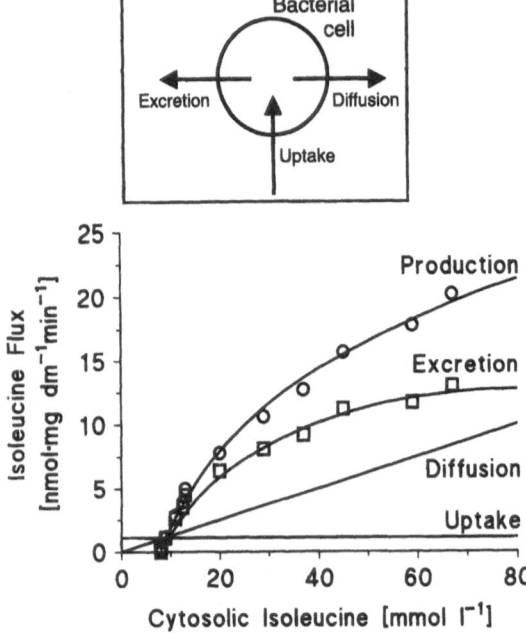

Fig. 10. Different fluxes of isoleucine across the membrane of *C. glutamicum* as observed during production of isoleucine [41]. The *lower part* of the figure shows the dependence of the three individual fluxes of carrier-mediated isoleucine uptake, carrier-mediated excretion and diffusion-controlled efflux on the internal isoleucine concentration under conditions of low external isoleucine. The individual fluxes, as shown in the *upper part* of the figure, sum up to isoleucine production according to their corresponding flux direction (i.e. according to their sign in Table 2). The direction of diffusion depends on the actual difference between the internal and external isoleucine concentration, and the same holds for the sum (production) as exemplified by the date of Table 2

Table 4. Contribution of the different transmembrane isoleucine fluxes in *C. glutamicum* and their regulation under different metabolic conditions

Isoleucine concentration (mmol l⁻¹)		State of activity of		Isoleucine flux[a] mediated by (nmol mg⁻¹ dw min⁻¹)			
internal	external	uptake system	excretion system	diffusion	uptake	excretion	sum = production
0.5	0.5	off	off	0	0	0	− 0
4	0.5	on	off	0.5	− 1.2	0	− 0.7
20	0.5	on	on	2.7	− 1.2	6	7.5
100	0.5	on	on	13	− 1.2	14	26
20	50	on	on	− 3.9	− 1.2	5.3	0
20	100	on	on	− 10.4	− 1.2	4.5	− 7.1
100	100	on	on	0	− 1.2	10	8.8
100	150	on	on	− 7	− 1.2	8	− 0.2

[a] Positive sign: efflux direction of flux; negative sign: uptake direction of flux

a massive energy-wasting futile cycle occurs via energy-dependent extrusion and diffusion-controlled reuptake of isoleucine. The same in principle occurs at lower concentrations with respect to isoleucine uptake and excretion, which are both energy-dependent (upper part of Table 4 and Fig. 10), although the uptake activity is relatively low and this futile cycle is thus not of high significance.

There are in fact indications that the combination of uptake and excretion systems may be of importance for a number of biotechnologically relevant processes. A scenario of this kind is indicated by recent findings on the excretion of aromatic amino acids by *C. glutamicum* [109]. It was shown that, during production of aromatic amino acids in a recombinant *C. glutamicum* strain, overexpression of the uptake system for aromatic amino acids which is also present in this organism decreased the yield of these amino acids as excreted by the producer strain. Consequently, it was found that elimination of the uptake activity increased the productivity. These findings can be interpreted as an indication of the presence of a futile cycle of aromatic amino acids, the elimination of which may lead to an increased yield. A similar interpretation was put forward to explain differences in citrate productivity by different strains of *Candida guillermondii* [110].

5 Modeling of Transport Processes

Despite early findings in fundamental studies on bacterial energy metabolism that uptake of nutrients consumes a significant part of the total energy for growth in the cell [20, 108], and despite the fact that the energy share of transport reactions may even be greater under situations of producing and non-growing cells, there are relatively few examples where uptake steps have been taken explicitly into account in models of cell metabolism. This of course holds even more for the process of product excretion. In the following, some examples of approaches to include essential transport reactions (nutrient uptake and product excretion) into models of cell metabolism will be discussed.

There are a number of different levels on which transport events can be included into concepts describing and modeling cell metabolism. When establishing a structured model, as has been done for several organisms [23], it is of course of principal importance to include the steps of uptake of the major carbon, energy and nitrogen sources into the setup of equations. In a very basic sense, i.e. without any further elucidation of the extent of their contribution, transport steps, and in particular that of nutrient uptake, have in fact been included formally in modeling procedures [99, 111, 112].

In descriptions based on analyses using the approach of the metabolic control theory, the focus lies in defining the contribution of every single step (control strength) to the overall control of the metabolism or a defined part of it (cf. [23]). Undoubtedly, transport steps may be important points of control in this respect and attempts have been made to quantify the share of these step in control strength by applying inhibitors, competitors or molecular techniques (recombinant strains). Valuable examples of metabolic control analysis have already been extended to transport steps in the case of mitochondrial metabolism in eukaryotic cells [73–75, 113, 114]. Interestingly, these investigations showed that the actual control strength of a given carrier system changes widely, e.g. from nearly zero to a value of more than 0.5, depending on the metabolic conditions due to regulatory mechanisms at the level of the functional interaction between the cytosolic and the mitochondrial compartment. This exemplifies the fact that, in general, the determination of the basic properties of a given carrier, i.e. the substrate affinity, the maximum capacity and the energy demand (mechanism) may be more or less useless for understanding metabolic fluxes in general, unless a detailed analysis of its significance is carried out under the particular metabolic conditions. Also, in the case of bacterial cells, uptake systems have been considered in control analysis, e.g. the lactose uptake system (lac-permease) of *E. coli* [115], or the phosphotransferase system (PTS) of sugar uptake [116]. Metabolic control analysis of glycolysis in yeast was extended to a detailed consideration of the reactions of glucose uptake [117]. The results of this study showed that glucose uptake is in fact the major rate-controlling step in yeast catabolism. With the increasing accessibility of carrier systems on the molecular level and thus to controlled variation of their activity, the models of metabolic flux control on the level of intermediary metabolism can be extended in the future by including transport steps.

As described in [63], establishment of dynamic models of the cell metabolism needs the measurement of time-dependent changes of metabolites by using fast sampling techniques. Provided methods are applied which include separation of the cell interior from the medium during or after sampling and quenching, transport steps can, in principle, be included in these type of analysis as has already been attempted, again for sugar uptake in yeast [27, 33]. In contrast to the situation in a static description of the metabolism, it again becomes obvious that the basic properties of a carrier system, as determined in resting cells or with isolated transport proteins, are only of limited value for dynamic models. The time-dependent changes of the external and internal concentration of the transported molecules within a time resolution suitable for resolving carrier activity kinetically must be analysed in detail in order to evaluate correctly the importance of a given carrier system for the control of the metabolic flux in the cell. This requirement critically limits the methods applicable to fast analysis of transport processes in cells in a defined metabolic state, e.g. when sampled from a bioreactor (see Sect. 3.4. and [63]).

In kinetic models of the cell metabolism (see [23]), transport steps are also important individual reactions. In general, as a first approach, the kinetics of transport systems is integrated into this type of model by inserting the Michaelis-Menten equation (see Sect. 3.3.), which in fact is frequently a satisfying approach. It is not the issue of this article to describe the kinetics in structured models of intermediary metabolism in general and of carrier-mediated transport in particular. Kinetic models of various transport systems, both from eukaryotic (e.g. the glucose facilitator, the erythrocyte anion carrier, the Na^+/glucose cotransporter, the Na^+, K^+ATPase, and the Ca^{2+}ATPase) and from prokaryotic cells (e.g. the lactose carrier, the maltose PTS, binding protein-dependent systems) have been established and described in numerous reviews and books (see for example [3, 7, 21, 38, 48, 118–123]). These data provide a large number of elaborated kinetic descriptions, which can then be integrated into structured kinetic models of metabolism, which are described in more detail in [23]. Models of this kind are helpful for the purpose of metabolic design from several aspects. When a mechanism is known in detail, quantitative evaluation of the limiting influence of cofactors (e.g. cotransported ions) and important functional parameters (e.g. electrochemical ion gradients or external pH) on transport function is possible. Furthermore, unless a full molecular analysis of the transport system(s) in a given organism for a particular substrate has been achieved, i.e by identification and isolation of the corresponding gene(s) which, frequently, is not the case in organisms of biotechnological relevance, only a functional (kinetic) analysis provides the data with respect to the possibility of the presence of a multiplicity of differently regulated systems (see Sect. 3.3). A practical example is the long-standing discussion on the possible multiplicity of glucose transport systems in *S. cerevisiae* which is of central significance in modeling the catabolic flux in this organism. The obvious discrepancies which exist regarding a correct kinetic analysis and the interpretation of the functional properties of sugar uptake [33, 124–127], will hopefully be solved by a detailed functional analysis of the properties of recombinant strains carrying only a single system each [128, M. Ciriacy, personal communication].

Transport processes can be incorporated into modeling approaches and finally into strategies of metabolic design in another aspect, namely the contribution of transport to the energetics of the cell or of a given part of the metabolism. It was estimated that transport contributes about 15–25% of the overall energy demand of the cell during growth [20, 108]. Contribution of substrate uptake and product release in energetic terms can in fact be significant, and not only with respect to the percentage of overall energy required for transport. Sugar molecules are frequently taken up in bacterial cells by the phospho*enol*pyruvate phosphotransferase systems (PTS) which stoichiometrically consume one molecule of phospho*enol*pyruvate (PEP) for one sugar molecule transported [21]. Thus, in the intermediary metabolism, this demand must be matched, which consequently leads to additional constraints in modeling (as an example see [111]) as well as to additional complications when, for example, products are studied, the synthesis of which includes PEP, such as aromatic amino acids [129]. Energetic considerations in flux modeling and especially in metabolic design are also important when alternative substrates can be used. *S. cerevisiae*, for example, possesses facilitator systems for glucose uptake, which do not need input of metabolic energy for their function, whereas maltose is taken up in cotransport with a proton [130, 131]. Since the ATP/H^+ stoichiometry of the plasma membrane H^+-ATPase in yeast is most likely 1, one molecule of ATP is used to take up one molecule of maltose, which is equivalent to 25% of the energy yield of the substrate [131]. Another instructive example is the comparison of glucose uptake in *E. coli* which is predominantly mediated by a PEP-consuming PTS system, whereas the same substrate is taken up in *Z. mobilis* by a facilitator system which does not need input of metabolic energy. The actual difference in the energy demand between these two setups is not significant, since in *Z. mobilis* the sugar molecule taken up has to be phosphorylated by a kinase in order to be channeled into the catabolic pathway. The response to a descreased or limiting sugar supply during growth, however, e.g. in terms of reversibility and efficiency of uptake, as well as the interelations with the intermediary metabolism are significantly different and have to be considered in detailed model (see also [63]). As an example, three different mechanisms of sugar uptake are compared in energetic terms in Fig. 11, namely a secondary, ion-coupled mechanism (A), a binding protein-dependent system (B) and a phospho*enol*pyruvate phosphotransferase system (C). Intuitively, the secondary system, simply coupled to the flux of one proton or sodium ion, seems to be the cheapest way, whereas the PTS system consuming the high-energy compound PEP is thought to have the highest energy demand. Just the opposite is true. In fact, the PTS system is the most economic way to take up carbohydrates if the consecutive connection to central metabolic pathways is taken into account, since no further activation by phosphorylation is necessary. Secondary systems need significantly more energy if coupled to the movement of other ions, since protons or sodium ions must first be actively transported out of the cell, consuming fractions of ATP or chemical energy of the respiratory chain substrate, depending on the actual stoichiometry (ion/ATP) of the particular ATPase or the H^+-extrusion stoichiometry of the respiratory chain. Only in the case of a facilitator mechanism of glucose uptake, as in *S. cerevisiae* or in *Z. mobilis*, the energy cost is more or less the same as compared to a PTS system (see above). Binding protein dependent systems (Fig. 11B) consume significantly more energy for substrate uptake, which, however may be necessary under the particular situation of extremely high accumulation ratios achieved by these systems (Sect. 2.1).

Although the importance of product excretion mechanisms has been recognized in several cases, e.g. that of aromatic amino acids in *E. coli* [132], of nucleotides in *Brevibacterium ammoniagenes* [133, 134], of citrate in various fungi [135, 136], and of various amino acids in *C. glutamicum* [1, 12, 40, 41, 83, 137, 138], they have mainly been neglected as individual steps in models so far, except for mentioning their importance in principle [e.g. 112] or incorporating them formally into equation systems, e.g. [139]. At least in the case of amino acid production the relevance of transport steps for a complete pattern of metabolic fluxes has been acknowledged recently [23, 99, 140]. Some essential basic information is now available in the particular case of lysine excretion in *C. glutamicum* for a meaningful implementation of this step into flux modeling. As described in Sect. 4.2, kinetic and energetic data have been obtained which led to a kinetic model of the carrier mechanism [40, 91]. It has been shown that lysine excretion is a critical step in the overall flux from glucose to external lysine [92, 93] and that the characteristics of this step, i.e. the properties of the lysine excretion carrier, differ between wild type and production strains [94]. Based on these data and on the kinetic model, an elaborated study using fed-batch and continuous cultivation of a lysine-producing *C. glutamicum* strain was carried out [141]. These investigations showed that (i) the lysine carrier activity of the bacterial cells in the reactor was modulated by the same parameters as pre-

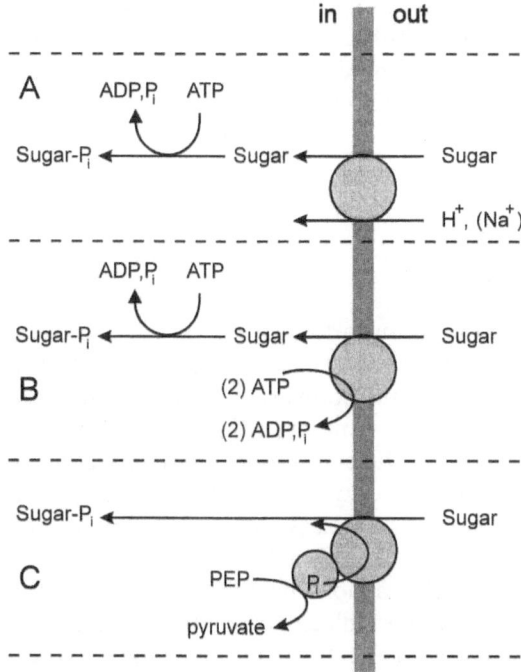

Fig. 11A–C. Energetic cost of different systems catalyzing sugar uptake: **A** a secondary, proton- (or Na$^+$-) coupled system; **B** a primary, binding protein-dependent system; **C** a phospho*enol*pyruvate (PEP) phosphotransferase system. The coupling stoichiometry of binding protein-depedent systems (molecules ATP hydrolyzed per molecule sugar transported) is not known exactly, but the most likely value is 2. For calculations see text

Table 5. Some relevant process parameters during fed-batch cultivation of a lysine producing strain of *C. glutamicum*

| Parameter | Relative value of parameter (%) at fermentation time | | | | | |
	10 h	20 h	30 h	40 h	50 h	60 h
μ	33	100	5	0	0	0
$Y_{X/S}$	100	88	53	30	13	0
$Y_{P/S}$	0	18	90	96	55	38
q_P	33	74	100	93	74	59
$[carrier]_{sp}$	57	97	100	86	69	57

The data are taken from [141], the specific carrier concentration was evaluated by kinetic experiments as described in [141]. The 100% values of the various parameters were: growth rate μ, 0.03 h^{-1}; biomass yield $Y_{X/S}$, 0.4 g dry weight g^{-1} glucose; product yield $Y_{P/S}$, 0.56 mol lysine mol^{-1} glucose; specific productivity q_P, 0.27 mmol lysine g^{-1} dry weight h^{-1}; specific carrier concentration $[carrier]_{sp}$, nmol g^{-1} dry weight

viously analysed using stationary cells [91, 95], e.g. the medium pH and the external lysine concentration, and (ii) the lysine productivity correlated closely with the specific carrier concentration, as estimated by kinetic experiments, which significantly changed in the course of growth, presumably due to regulation on the level of expression (Table 5). The molecular identification of the lysine excretion carrier [39, M. Vrljic and L. Eggeling, unpublished results] and the accessibility for modification and controlled expression will help in providing quantitative data for further progress in modeling the metabolic situation in *C. glutamicum* under conditions of lysine production.

6 Conclusions and Perspectives

In this contribution, it was argued that transport reactions, i.e. both uptake of nutrients or supplements and excretion of products, may be important factors in the control of cell metabolism and may thus be relevant for biotechnological applications and for metabolic engineering. A number of problems have been discussed which arise when trying to include transport steps properly, e.g. comparable to that of steps catalyzed by enzymes, in structured models of metabolism which are intended to be used as a basis for metabolic engineering purposes. These problems are basically due to the fact that transport systems are nowhere near as well studied as enzymatic reactions, mainly because membrane-embedded carrier proteins are much more difficult to handle experimentally compared to soluble enzymes. Similarly, the methods for analysing a carrier's actual activity, for example in a reactor, are significantly more complicated than that of an enzyme. In enzymatic reactions substrate and product are different and their interconversion can thus be followed, whereas in the case of carrier systems in general the substrate only changes its location, and consequently a two-compartment system is necessary for analysis. Transport steps are therefore frequently inadequately treated and included in models, since a lot of essential knowledge is still missing with respect to their functional properties. Mostly literature data are used, which are obtained by measurements of isolated and washed cells, i.e. under very different conditions not easily transferable to the situation in a bioreactor, both with respect to growth and production conditions.

Another restriction in the accessibility of transport systems to metabolic engineering is their relatively difficult access for molecular biology techniques. Controlled overexpression of membrane proteins in recombinant strains is frequently difficult to obtain. Knowledge and expertise in this field, however, is rapidly increasing which may also open this approach for a broader application. In view of the fact that only a very limited number of examples have been studied in detail, the observation that transport may in fact be a rate-limiting step in a number of biotechnologically relevant cases argues for a broader survey in this respect and for a broader consideration in concepts of flux modeling and metabolic engineering. Particular examples of purposeful engineering in order to increase the spectrum of substrates and products as well as the yield and the productivity of certain processes on the basis of the results discussed in this article are obvious. Examples are (i) the release of limiting steps (points of major control) which may be on the level of nutrient uptake and/or on the level of product excretion, (ii) the modification of the availability of the kind of uptake or excretion systems or their regulation adapted to the metabolic conditions of a particular organism, as well as (iii) the elimination of counter-productive combinations of transport reactions such as the presence of a multiplicity of systems acting in different directions, or (iv) the modulation (up or down) of classic energetic futile cycles.

For a successful approach, we obviously need more detailed knowledge of the consequences of different transport mechanisms for different purposes in terms of biotechnology. Although this seems to be a trivial question at first sight, this article should have made clear that it is in fact not trivial to decide on a purely theoretical basis, for example, whether a facilitator system for nutrient uptake, which does not need the input of metabolic energy but cannot catalyse substrate accumulation, is advantageous under particular conditions of fermentation compared to an energy-dependent and thus energy-consuming uptake system which concentrates nutrients from the surroundings. More physiological, biochemical and molecular data are necessary to answer this type of important question which starts to become interesting for biotechnology in view of the increasing availability and use of recombinant cells designed by metabolic engineering.

Hopefully, a more detailed functional analysis of transport systems, especially under conditions comparing growth and production, will provide a more elaborate and reliable way to include transport steps in metabolic models. This in turn will lead to more detailed proposals for engineering both the intermediary metabolism as well as transport steps. Finally, combining these strategies with improved fine-tuning of process parameters, according to particular demands of the cell metabolism including modulation of transport systems (uptake and excretion), opens the perspective of true molecular approaches of metabolic engineering with respect to transport.

7 References

1. Krämer R (1994) FEMS Microbiol Rev 13: 75
2. Mitchell P (1967) In: Florkin M, Stotz EH (eds) Comprehensive biochemistry, Vol 22, Elsevier, Amsterdam, p 167
3. Stein WD (1986) Transport and diffusion across cell membranes, Academic Press, New York
4. Harold F (1986) The vital force: a study of bioenergetics, Freeman, New York
5. Konings WN, Poolman B, Driessen AJM (1992) FEMS Microb Rev 88: 93
6. Poolman B, Konings WN (1993) Biochim Biophys Acta 1183: 5
7. Krämer R (1994) Biochim Biophys Acta 1185: 1
8. Läuger P (1985) Biophys J 47: 581
9. Eisenberg RS (1990) J Membrane Biol 115: 1
10. Heinz E (1978) Mechanics and energetics of biological transport, Springer, Berlin
11. Krupka RM (1993) Biochim Biophys Acta 1183: 105
12. Krämer R (1994) Arch Microbiol 162: 1

13. Shiio I (1986) In: Aida I, Chibata K, Nakayama K, Takinami K, Yamada H (eds) Biotechnology of Amino Acid Production, Kodansha, Tokyo/Elsevier, Amsterdam, p 188
14. Katsumata R, Ikeda M (1993) Bio-Technology 11: 921
15. Doelle HW, Kirk L, Crittenden R, Toh H, Doelle MB (1993) Crit Rev Biotechnol 13: 57
16. Anantharam V, Allison MJ, Maloney PC (1989) J Biol Chem 264: 7244
17. Jencks WP (1980) Adv Enzymol. 51: 75
18. Jencks WP (1989) Meth Enzymol 171: 145
19. Krupka RM (1989) J Membrane Biol 109: 151
20. Stouthamer AH (1973) Anton, Leeuwenhoek Int J Gen M 39: 545
21. Postma PW, Lengeler JW, Jacobson GR (1993) Microbiol Rev 57: 543
22. Lengeler JW (1993) Anton Leeuwenhoek Int J Gen M 63: 275
23. Eggeling L, Sahm H, de Graaf AA (1995) Adv Biotechnol, this volume, p.
24. Klingenberg M, Pfaff E (1967) Meth Enzymol 10: 680
25. Rottenberg H, (1978) In: Azzone GF (ed) The proton and calcium pumps, Elsevier, Amsterdam, p 125
26. Bröer S, Krämer R (1990) J Bacteriol 172: 7241
27. de Koning W, van Dam K (1992) Anal Biochem 204: 118
28. Kirk K (1990) NMR in Biomedicine 3: 1
29. Axe DD, Bailey JE (1994) Biotechnol. Bioeng 43: 242
30. Kuchel PW, Kirk K, King GF (1995) In: Hilderson HJ, Ralston GB (eds) physiochemical methods in the study of biomembranes, Plenum Press, London, p 247
31. Schoberth SM, de Graaf AA (1993) Anal. Biochem. 210: 123
32. Schoberth SM, Chapman BE, Kuchel PW Wittig R, Grotendorst J, Jansen P, de Graaf AA (1994) In: Ingman LP, Jokisaarie J, Lounila J (eds) 12th European Experimental NMR Conf., Oulu, p 43
33. Walsh MC, Smits HP, Scholte M, van Dam K (1994) J Bacteriol 176: 953
34. Fath MJ, Kolter R (1993) Microbiol Rev 57: 995
35. Lewis K (1994) Trends Biochem Sci 19: 119
36. Midgley M (1986) J Gen Microbiol 132: 3187
37. Ruhrmann J, Sprenger GA, Krämer R (1994) Biochem Biophys Acta 1196: 14
38. Lieb WR (1982) In: Ellory JC, Young JD (eds) Red cell membranes: a methodological approach, Academic Press, London, p 135
39. Vrlijc M, Kronemeyer W, Sahm H, Eggeling L (1995) J Bacteriol 177:
40. Bröer S, Krämer R (1991) Eur J Biochem 202: 131
41. Zittrich S, Krämer R (1994) J Bacteriol 176: 6892
42. Segel IH (1975) Enzyme kinetics, Wiley & Sons, New York
43. Cleland WW (1963) Biochim Biophys Acta 67: 104
44. Cleland WW (1970) In: Boyer P (ed) The Enzymes, Vol. 2, Academic Press, p 1
45. Läuger P. (1980) J Membrane Biol. 57: 163
46. Stein WD (1989) Meth Enzymol 171: 23
47. Deves R, Krupka RM (1989) Meth Enzymol 171: 113
48. Krupka RM (1989) Biochem J 260: 885
49. Stein WD (1990) Channels, carriers and pumps, Academic Press, San Diego
50. Läuger P (1991) Electrogenic Ion Pumps, Sinauer Associates, Sunderland
51. Benz R (1988) Ann Rev Microbiol 42: 359
52. Carruthers A (1990) Physiol Rev 70: 1135
53. Bisson LF, Coons DM, Kruckeberg AL, Lewis DA (1993) Crit Rev Biochem Molec Biol 228: 259
54. Gould GW, Holman GD (1993) Biochem J 295: 329
55. Mueckler M (1994) Eur J Biochem 219: 713
56. Poolman B, Driessen AJM, Konings WN (1987) Microbiol Rev 51: 498
57. Krämer R, Lambert C, Ebbighausen H, Hoischen C (1990) Eur J Biochem 194: 929
58. Padan E, Zilberstein D, Schuldiner S (1981) Biochim Biophys Acta 650: 151
59. Booth IR (1985) Microbiol Rev 49: 359
60. Baronofsky JJ, Schreurs WJA, Kashket ER (1984) Appl Environ Microbiol 48: 1134
61. Herrero AA, Gomez RF, Snedecor B, Tolman CJ, Roberts MF (1985) Appl Microbiol Biotechnol 22: 53
62. Weusthuis RA, Pronk JT, van den Broek PJA, van Dijken JP (1994) Microbiol Rev 58: 616
63. Weuster-Botz D, de Graaf AA (1995) Adv Biotechnol, this volume, p.

64. Bisson LF, Fraenkel DG (1984) J Bacteriol 159: 1013
65. Röhm S, Hermann S, Röhm KH, Fuhrmann GF (1992) J Biotechnol 27: 85
66. Ter Kuile BH, Cook M (1994) Biochim Biophys Acta 1193: 235
67. Kacser H, Burns JA (1973) Symp Soc Exp Biol 27: 65
68. Heinrich R, Rapoport TA (1974) Eur J Biochem 42: 89
69. Westerhoff HV, van Dam K (1987) Thermodynamics and control of free energy transduction, Elsevier, Amsterdam
70. Westerhoff HV, Kell DB (1987) Biotechnol Bioeng 30: 101
71. Small JR, Kacser H (1993) Eur J Biochem 213: 613
72. van Dam K, Jansen N, Postma P, Richard P, Ruijter G, Rutgers M, Smits HP, Teusink B, van der Vlag J, Walsh M, Westerhoff HV (1993) Anton Leeuwenhoek Int J Gen M 63: 315
73. Groen AK, Wanders RJA, Westerhoff HV, van der Meer R, Tager JM (1982) J Biol Chem 257: 2754
74. Tager JM, Wanders RJA, Groen AK, Kunz W, Bohnensack R, Küster R, Letko G, Böhme G, Duszinski J, Woijtczak L (1983) FEBS Lett 151: 1
75. Westerhoff HW, Plomp PJAM, Groen AK, Wanders RJA, Bode JA, van Dam K (1987) Arch Biochem Biophys 257: 154
76. Romano AH (1986) Trends Biotech 4: 207
77. Poolman B (1990) Mol. Microbiol 4: 1629
78. DiMarco AA, Romano AH (1985) Appl Environ Microbiol 49: 151
79. Parker C, Barnell WO, Snoep JL, Ingram LO, Conway T (1995) Mol Microbiol 15: 795
80. Weisser P, Krämer R, Sahm H, Sprenger GA (1995) J Bacteriol 177: 3351
81. Neijssel OM, Tempest DW (1979) Symp Soc Gen Microbiol 29: 53
82. Tempest DW, Neijssel OM (1992) FEMS Microbiol Lett 100: 169
82a. Sonnleitner B, Käppeli O (1993) Biotechnol Bioeng 28: 927
82b. Sonnleitner B, Hahnemann U (1994) J Biotechnol 38: 63
83. Gutmann M, Hoischen C, Krämer R (1992) Biochim Biophys Acta 1112: 115
84. Russell JB, Cook GM (1995) Microbiol Rev 59: 48
85. Higgins CF (1992) Ann Rev Cell Biol 8: 67
86. Doige CA, Ames GFL (1993) Annu Rev Microbiol 47: 291
87. Payne JW, Nisbet TM (1980) FEBS Lett 119: 73
88. Matin A, Konings WN (1973) Eur J Biochem 34: 58
89. Driessen AJM, Konings WN (1990) In: Krulwich TA (ed) Bacterial Energetics, Academic Press, London, p 449
90. Krämer R, Boles E, Eggeling L, Erdmann A, Gutmann M, Kronemeyer W, Palmieri L, Zittrich S (1994) Biochim Biophys Acta 1187: 245
91. Bröer S, Krämer R (1991) Eur J Biochem 202: 137
92. Schrumpf B, Eggeling L, Sahm H (1992) Appl Microbiol Biotechnol 37: 566
93. Schrumpf B, Sahm H, Eggeling L (1995) submitted for publication
94. Bröer S, Eggeling L, Krämer R (1993) Appl Environm Microbiol 59: 316
95. Erdmann A, Weil B, Krämer R (1994) Appl Microbiol Biotechnol 42: 604
96. Linton JD (1990) FEMS Microbiol Rev 75: 1
97. Kiss RD, Stephanopoulos G (1991) Biotechnol Progr 7: 501
98. Coello N, Pan JG, Lebault JM (1992) Appl Microbiol Biotechnol 38: 259
99. De Hollander JA (1994) Appl Environ Microbiol 42: 508
100. Buurman ET, Demattos MJT, Neijssel OM (1991) Arch Microbiol 155: 391
101. Mulder MM, Teixeira MJ, Postma PW, van Dam K. (1986) Biochim Biophys Acta 851: 223
102. Stouthamer AH (1979) In: Quayle JR (ed) International Review of Biochemistry, Vol 21, University Park Press, Baltimore, p 1
103. Tempest DW, Neijssel OM (1984) Annu Rev Microbiol 38: 459
104. Verseveld H, Chresbo W, Braster M, Stouthamer A (1984) Arch Microbiol 137: 176
105. Kleiner D (1985) FEMS Microbiol Rev 32: 87
106. Verdoni N, Aon MA, Lebeault JM, Thomas D (1990) J Bacteriol 172: 6673
107. Chao YP, Liao JC (1994) J Biol Chem 269: 5122
108. Stouthamer AH, van Verseveld HW (1985) In: Bull AT, Dalton H (eds) Comprehensive biotechnology; The principles, applications and regulations of biotechnology in industry, agriculture and medicine, Pergamon Press, Oxford, p 215
109. Ikeda M, Katsumata, R (1994) J Ferment Bioeng 78: 420
110. Gutierrez NA, Maddox IS (1993) Appl Microbiol Biotechnol 39: 604

111. Vallino JJ, Stephanopoulos G (1993) Biotechnol Bioeng 41: 633
112. Varma A, Boesch BW, Palsson PO (1993) Biotech Bioeng 42: 59
113. Mazat JP, Jean-Bart E, Rigoulet M, Guerin B (1986) Biochim Biophys Acta 849: 7
114. Gellerich FN, Kunz WS, Bohnensack R (1990) FEBS Lett 274: 167
115. Dykhuizen DE, Dean AM, Hartl DE (1987) Genetics 115: 25
116. van der Vlag J, van t'Hof R, van Dam K, Postma PW (1995) Eur J Biochem 230: 170
117. Cortassa S, Aon MA (1994) Enzyme Microb Technol 16: 761
118. Lieb WR, Stein WD (1974) Biochim Biophys Acta 373: 178
119. Page MGP, West IC (1981) Biochem J 196: 721
120. Carruthers A (1991) Biochemistry 30: 3898
121. Lolkema JS, Carrasco N, Kaback HR (1991) Biochemistry 30: 1284
122. Passow H (1992) In: Bamberg E, Passow H (eds) Progress in Cell Research, Vol. 2, Elsevier, Amsterdam, p 1
123. Krämer R, Palmieri F (1992) In: Ernster L (ed) Molecular mechanisms in bioenergetics, Elsevier, Amsterdam, p 359
124. Bisson LF, Fraenkel DG (1983) Proc Natl Acad Sci USA 80: 1730
125. Wrede C, Völker B, Küntzel H, Fuhrmann GF (1992) J. Biotechnol. 27: 47
126. Coons DM, Boulton RB, Bisson LF (1995) J Bacteriol 177: 3251
127. Does AL, Bisson LF (1989) J Bacteriol 171: 1303
128. Reifenberger E, Freidel K, Ciriacy M (1995) Mol Microbiol 16: 157
129. Patnaik R, Liao JC (1994) Appl Environ Microbiol 60: 3903
130. Weusthuis RA, Adams H, Scheffers WA, van Dijken JP (1993) Appl Environ Microbiol 59: 3102
132. Brown KD (1971) J Bacteriol 106: 70
133. Teshiba S, Furuya A (1983) Agric Biol Chem 47: 2357
134. Teshiba S, Furuya A (1984) Agric Biol Chem 48: 1311
135. Kubicek CP, Rohr M (1986) CRC Crit Rev Biotechnol 3: 331
136. Burgstaller W, Zanella A, Schinner F (1994) Arch Microbiol 161: 75
137. Milner JL, Vink B, Wood JM (1987) CRC Crit Rev Biotechn 5: 1
138. Hoischen C, Krämer R (1989) Arch Microbiol 151: 342
139. Drozdov-Tikhomirov LN, Skurida GI (1977) Mol Biol (USSR) 11: 653
140. Marx A, de Graaf AA, Wiechert W, Eggeling L, Sahm H (1995) submitted for publication
141. Kelle R, Laufer B, Brunzema C, Weuster-Botz D, Krämer R, Wandrey C (1995) Biotechnol. Bioengin., in press

Reaction Engineering Methods to Study Intracellular Metabolite Concentrations

D. Weuster-Botz, A.A. de Graaf

Institute of Biotechnology, Research Center Jülich, D-52425 Jülich, Germany

The analysis of intracellular metabolite concentrations is of basic importance for metabolic engineering of microorganisms. In vivo NMR-spectroscopy as a non-invasive technique to measure intracellular metabolite concentrations and rapid sampling devices as invasive techniques are reviewed. The methods are discussed from a reaction engineering point of view. The objective is to obtain intracellular concentration data under well defined physiological conditions in balanced steady state and defined transitional states as well. Application examples are given for a membrane-cyclone-reactor configuration designed to achieve high signal sensitivity with in vivo ^{31}P-NMR and ^{13}C-NMR spectroscopy as well as for a sampling tube device designed for high sampling rates ($2\ s^{-1}$). This sampling device enables the measurement of dynamic metabolite profiles at a time scale of a few seconds.

Advances in Biochemical Engineering
Biotechnology, Vol. 54
Managing Editor: T. Scheper
© Springer-Verlag Berlin Heidelberg 1996

List of Symbols and Abbreviations

NMR	nuclear magnetic resonance
T_r	length of NMR scan (s)
SNR	signal-to-noise-ratio
S(t)	frequency component of the NMR signal
α	flip angle
T_1	spin-lattice relaxation time, characterizing the rate at which the magnetic polarization of exited nuclear spins returns to its equilibrium (s)
T_2^*	effective transverse relaxation time, characterizing the rate of decay of the observed NMR-signal (s)
r.f.	radio frequency
HFBR	hollow fiber bioreactor system
K	ratio of mean residence time of the cells to the mean residence time of medium
k_La	volumetric oxygen transport coefficient (s^{-1})
x(t)	tracer input function
y(t)	tracer response function
g(t)	weighting function
l(t)	metabolite input function
m(t)	metabolite response function
X(s)	Fourier transform of x(t)
Y(s)	Fourier transform of y(t)
G(s)	Fourier transform of g(t)
L(s)	Fourier transform of l(t)
M(s)	Fourier transform of m(t)
AMP	adenosine-monophosphate
ADP	adenosine-diphosphate
ATP	adenosine-triphosphate
NDP	nucleoside-diphosphate
NTP	nucleoside-triphosphate
NAD(H)	nicotinamide-adenine-dinucleotide
NADP(H)	nicotinamide-adenine-dinucleotide-phosphate
SP	sugar-phosphate
P_{in}	intracellular phosphate
P_{ex}	extracellular phosphate
UDP-sugar	uridinediphosphate-sugar
UDPG	uridinediphosphate-glucose
G6P	glucose-6-phosphate
F6P	fructose-6-phosphate
DHAP	dihydroxyacetone-phosphate
FBP	fructose-1,6-bisphosphate
PEP	phosphoenolpyruvate
6PG	6-phospho-gluconate
GAP	glyceraldehyde-3-phosphate

1 Introduction

Intracellular metabolite concentrations play important regulatory roles in the cellular metabolic network of microorganisms. Together with information about kinetic properties of the enzymes involved, knowledge of the in vivo concentrations of intermediary metabolites are of fundamental importance for the characterization of the microbial metabolism by dynamic modeling. Dynamic models of (parts of) the microbial metabolism are an important tool for predicting the effects of genetic modifications. Thus the measurement of in vivo metabolite concentrations is of basic importance for metabolic engineering of microorganisms.

Techniques for in vivo metabolite analysis are divided into two different approaches: non-invasive and invasive techniques. Non-invasive techniques like nuclear magnetic resonance (NMR) spectroscopy have the advantage that a view into living cells is possible without any disturbance of the metabolism of the cells [1]. Other non-invasive techniques like NADH fluorescence [2] lack the ability to obtain intracellular concentrations of multiple metabolites at the same time. As NMR spectroscopy has been exhaustively used to characterize the elementary biochemical processes in living cells [3–5], part of this review deals with the application of in vivo NMR to non-invasive analysis of intracellular metabolites in microorganisms.

Invasive techniques for in vivo metabolite analysis must guarantee that the measured metabolite concentration in the sample represents the real concentration in the cells. If this requirement can be fulfilled, then the invasive techniques have the advantage that for each metabolite the best available analysis method can be applied. The use of rapid sampling and inactivation techniques for invasive metabolite analysis is therefore another subject dealt with in this review.

An elementary requirement for the identification of metabolic models by measuring the intra- and extracellular metabolite concentrations is that the microorganisms are in a defined metabolic state. Therefore continuous cultivation techniques in controlled bioreactors are necessary to achieve balanced steady state conditions of a microbial system. Unfortunately, the cell concentrations in stirred tank reactors commonly used for continuous steady state cultivations are low. As intracellular concentrations of important metabolites range down to the $\mu mol\, l^{-1}$ scale and the cytosolic volume is about 1% of the total volume in the reactor due to the low cell concentration, highly sensitive analytical methods have to be applied.

Dynamic investigations require a balanced steady state as well, because dynamic models approximate metabolism around a certain stationary state. The complexity of dynamic modeling of microbial metabolism can be reduced if regulation of DNA level can be ignored at a certain steady state (growth rate). This is possible only if dynamic experiments can be monitored on a time scale smaller than the time constants for changes in intracellular enzyme concentrations. These minimal time constants are assumed to be higher than a few hundreds of seconds [6].

As may have become clear, the basis for studying intracellular metabolite concentrations is constituted by reaction engineering methods which guarantee cultivation of microorganisms in a balanced steady state and in a defined transitional state as well, and fulfill at the same time the requirements of the applied analytical technique to measure in vivo metabolite concentrations. For this reason the primary focus of this contribution will be, on the one hand, on reaction engineering methods for in vivo NMR-studies and, on the other hand, on reaction engineering methods for rapid sampling. Application examples will be given for both approaches.

2 Reaction Engineering for In Vivo NMR-Studies

2.1 Principles of In Vivo NMR

NMR (Nuclear Magnetic Resonance) was discovered in 1946 simultaneously by Bloch et al. [7] and Purcell et al. [8]. The physical principle of this phenomenon is that nuclei that possess a non-zero spin, when brought into a strong static magnetic field after being excited, can emit oscillating magnetic fields that may be detected by a radiofrequency detection coil system. The frequency of those oscillations varies linearly with the applied magnetic field strength and is proportional to the nucleus-specific gyromagnetic ratio. Furthermore, identical nuclei present in chemically different environments (i.e., being more or less strongly shielded from the magnetic field by the local electron density) produce signals with different frequencies [9]. Thus, the three different proton groups of ethanol can emit three well-defined different NMR signals. The signal intensity is directly proportional to the amount of contributing nuclei. Due to this characteristic, NMR spectroscopy rapidly became established as a spectroscopic tool for the identification and quantitation of compounds dissolved in liquids. Then, with the advent of so-called two-dimensional measurement techniques, NMR evolved into the invaluable tool for structure elucidation of chemically or biologically synthesized molecules it is nowadays.

It was realized relatively early that NMR could also be used to characterize metabolic processes going on in living cells. Pioneering work showed that using ^{31}P NMR, intracellular metabolite concentrations of energy-rich compounds such as ATP, ADP, sugar phosphates, and NAD(P) (H), as well as intracellular pH, can be measured in vivo [10,11]. Using the NMR-sensitive stable carbon isotope ^{13}C as a tracer (its natural abundance is only 1.1%), it was possible to follow the fate of individual carbon atoms in a sequence of metabolic reactions in vivo [12]. These features have been exhaustively used to characterize the elementary biochemical processes in the living cell. The studies of Shulman et al. [3], Den Hollander et al. [4] and Ugurbil et al. [5] represent classical examples in this respect. An excellent introduction to in vivo NMR covering methodological, instrumental and physiological, as well as practical aspects, has been written by Gadian [1].

An important drawback of NMR, however, is its inherent low sensitivity when observing biologically relevant nuclei like ^{31}P, ^{13}C and [15]N. Since protons have a much higher NMR receptivity, the use of ^{1}H NMR in vivo was also attempted [12] but appeared to be severely limited by the narrow spectral range (causing excessive overlap of signals) and the presence of an overwhelming water signal up to 10 000 times stronger than the metabolite signals of interest [13]. Therefore, the first and the majority of the later ^{31}P, ^{13}C and ^{15}N NMR experiments on cell suspensions were conducted with very high cell densities in the samples, the ratio of cell wet weight to the total sample volume in most cases being in the range of 10–50%. In these studies, massive difficulties were encountered with respect to adequate supply of nutrients and oxygen as well as accumulation of toxic metabolic end products. This especially applies to microorganisms, which consume oxygen at specific rates up to 30 times faster than mammalian cell cultures. Consequently, in the simple NMR tube, it is in general impossible to maintain a steady state over periods longer than the few minutes required for a single measurement. Over the years, various solutions have been published that were sought to circumvent these problems. They are briefly reviewed and discussed below.

For a better understanding of the strategies used for the design of the various in situ cultivation systems, the principal parameters governing NMR signal intensity and resolution should be briefly introduced first. A typical NMR experiment uses a series of equally spaced scans, each scan having a length T_r (typically 500–5000 ms) and consisting of a short radiofrequency pulse (typically 10–50 μs) for excitation followed by data acquisition (typically 50–1000 ms) and a subsequent delay time for repolarization of the nuclear spins. During data acquisition, the NMR signal is digitized and stored in computer memory. The acquired NMR signals from all scans are added on-line point-by-point to finally produce the total NMR signal acquired during the experiment. Because the total signal increases linearly with the number of scans whereas the noise increases with the square root of the number of scans, in NMR the signal-to-noise ratio (SNR) increases with the square root of the number of scans, i.e., also only with the square root of the measurement time.

The time envelope and amplitude proportionality of a single frequency component S(t) of the NMR signal thus obtained from a sample present in an NMR tube, under steady-state pulsing at a repetition rate of $1/T_r (s^{-1})$ with flip angle α, and further characterized by relaxation time constants T_1 and T_2^* (explained below), is given [14] by

$$S(t) = c[(1 - \exp[-T_r/T_1])\sin(\alpha)/(1 - \cos(\alpha)\exp[-T_r/T_1])]$$
$$\times \exp[-t/T_2^*], \tag{1}$$

with c = proportionality constant. T_1 is the spin-lattice, or longitudinal, relaxation time that characterises the rate at which the magnetic polarization of excited nuclear spins returns back to its equilibrium. T_2^* is the effective transverse relaxation time that characterises the rate of decay of the observed NMR-signal. Under in vivo conditions, in many cases (especially with ^{31}P) $T_2^* \ll T_1$ [1]. In practice, the signal is sampled over a time span of about $3T_2^*$ and, after Fourier transformation of the signal, the line width of the resonance in the spectrum is about $1/3T_2^*$ (Hz) [1]. For fully relaxed conditions, $\alpha = 90°$ is used and $T_r > 5T_1$ to ensure complete repolarization of the nuclear spins before the next scan starts. This gives maximum NMR signals, although collected at a very low rate.

Equation (1) can be used to adjust the experimentally variable parameters α and T_r for signal detection with optimal sensitivity, according to the actual values T_2^* and T_1 in the sample. Obviously, at fixed α the NMR signal amplitude decreases with increasing T_1 and/or increasing pulse repetition rates due to so-called saturation. This is simply because the nuclear spins are given progressively less time to repolarize between successive scans. The clue for getting a higher sensitivity (defined as SNR per unit time) is using values of α smaller than 90°. It can be shown [14] that optimal sensitivity is obtained when α and T_r are matched such that $\cos(\alpha) = \exp[-T_r/T_1])$ in Eq. (1), with the constraint that $T_r <$ approx. $0.5T_1$.

2.2 Simple In Vivo NMR-Systems: Aerated Cell Suspension in the NMR-Tube

To oxygenate dense cell suspensions in the NMR tube (10–20 mm diameter, 5–20 ml of sample), in the Shulman group a device was developed known as the double bubbler apparatus [15]. One set of bubblers (diameter approx. 100 μm) was situated at the bottom of the yeast cell suspension to provide some oxygenation and to effect a stirring of the suspension that prevented the cells from settling. A set of larger diameter (approx. 500 μm) bubblers was positioned 1–2 cm above the NMR detection coil to provide most of the oxygen without impairing the homogeneity of the cell suspension in the detection region. Since the magnetic susceptibility of cell cultivation medium differs strongly from that of the gas bubbles (oxygen being paramagnetic), the presence of larger bubbles has a deleterious effect on the magnetic field homogeneity, which causes a strong broadening of the peaks in the NMR spectrum. The double bubbler system avoids such larger gas bubbles, and at the same time stabilises pH by removing volatile CO_2 accumulating in the cell suspension during substrate catabolism. The apparatus was reported to allow sufficient oxygenation of suspensions of yeast of 20% wet weight per volume, albeit at a temperature of only 20 °C to slow down metabolism [16, 17].

The group of Santos introduced an alternative method of mixing and oxygenating cell suspensions in the NMR tube while avoiding the presence of bubbles in the detection volume [18]. Their system can be considered as a miniaturization of the familiar air lift bioreactor (Fig. 1). In an outer conventional 10 mm

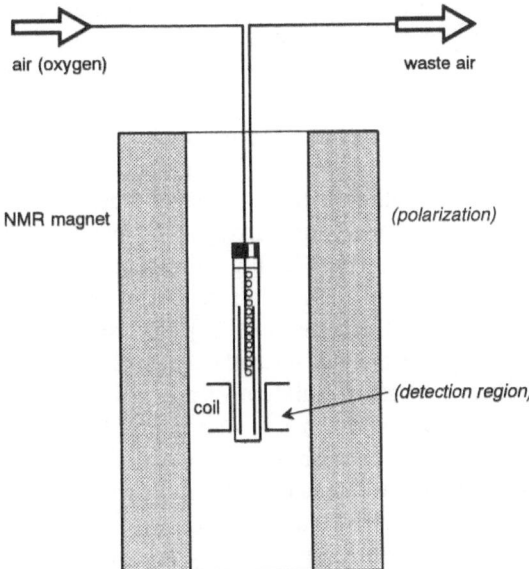

Fig. 1. Principle of an air lift probe for oxygenation and mixing cell suspensions in an NMR spectrometer

NMR tube, a second 5 mm diameter tube is concentrically positioned such that it extends from approx. 5 mm above the bottom of the outer tube to approx. 5 cm above the detection volume. A capillary is inserted in the central tube and ends approx. 1 cm above the detection volume. A stream of bubbles rising from the capillary drags the cell suspension through the inner tube; it recirculates down through the annulus. A subsequent study [19] of the mass transfer characteristics of this device indicated that it is well suited to oxygenate a suspension of about $10 \, \text{g} \, \text{l}^{-1}$ dry weight (i.e., about 6% wet weight per volume) consuming oxygen at a specific rate qO_2 of $0.7 \, \text{mmol} \, \text{g}^{-1} \, \text{h}^{-1}$, the $k_L a$-value for oxygen transfer of the system being $0.008 \, \text{s}^{-1}$. However, in the literature values of qO_2 as high as $20 \, \text{mmol} \, \text{g}^{-1} \, \text{h}^{-1}$ have been reported for microorganisms. Therefore both methods of oxygenating cell suspensions in the NMR tube are suspect to generating conditions that are not truly aerobic with the very high cell densities commonly used. Moreover, it is highly questionable if the physiological state of cells in these dense suspensions is representative of their state in diluted suspension during normal growth in a bioreactor.

2.3 Perfusion Reactor Systems: Control of Reaction Conditions

Perfusion systems have the advantage that a constant supply of substrate can be maintained while metabolic products are simultaneously removed. The first perfusion system for NMR studies was described by Ugurbil et al. [20]. It was used to study anchorage-dependent cells attached to microcarrier beads. Subsequently, first systems for hollow fiber perfusion [21] as well as embedding of cells in agarose-threads [22] were reported. The thread technique has become very popular because of its ease and applicability to a wide variety of cells and NMR spectrometer configurations. However, the cells are in partly uncontrolled conditions during preparation of the system, and the technique has a high probability of establishing perfusion heterogenities. Moreover, metabolic characteristics of immobilized cells may be different from suspended cells as for instance shown in a study of alginate-entrapped *Saccharomyces cerevisiae* [23]. At present, hollow fiber bioreactor systems (HFBR, Fig. 2) appear to be the optimal systems for the successful long-term cultivation in situ in the NMR magnet of mammalian cells at very high cell densities [24–26]. Best results thus far were obtained with a system equipped with microporous (0.2 μm) acetate/cellulose nitrate membrane fibers that allow bulk convective flow throughout the extracapillary space [26]. Using these systems, in vivo ^{31}P NMR spectra with unsurpassed resolution and SNR have been obtained and C-6 glioma cells could be held in a viable state at density 4×10^8 cells ml^{-1} for more than 1000 h. Anchorage-dependent cells could also be cultivated. However, with increasing cell densities, steady state gradients in nutrients, especially O_2, are still likely to occur [27].

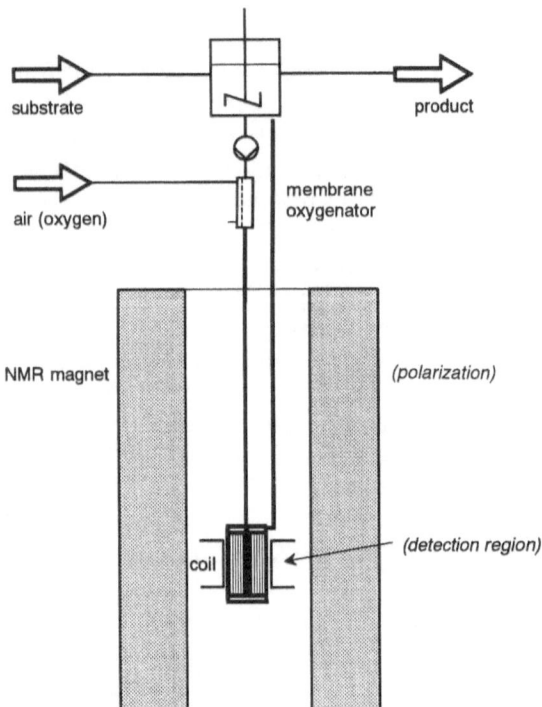

Fig. 2. Principle of a hollow fiber system for long-term cultivation of mammalian cells in an NMR spectrometer

While the HFBR nowadays are near-perfect for the cultivation of mammalian cells, they appeared much less suited for fast-growing microorganisms, especially with very high oxygen demands, since biomass cannot be controlled and high transport rates lead to non-uniform environments due to the mass transfer resistance within fibers and packed cell masses [28]. Thus, a HFBR study with *Escherichia coli* [29] clearly showed severe mass transfer limitations even under anaerobic conditions. Cells in high densities could only adequately be supplied with nutrients after T had been lowered to 16 °C [29].

2.4 Suspended Cell Reactor System with On Line NMR

From the above it is clear that microorganisms present special problems regarding adequate supply of nutrients because of their very high specific consumption rates. On the other hand, cultivation of these organisms in dilute

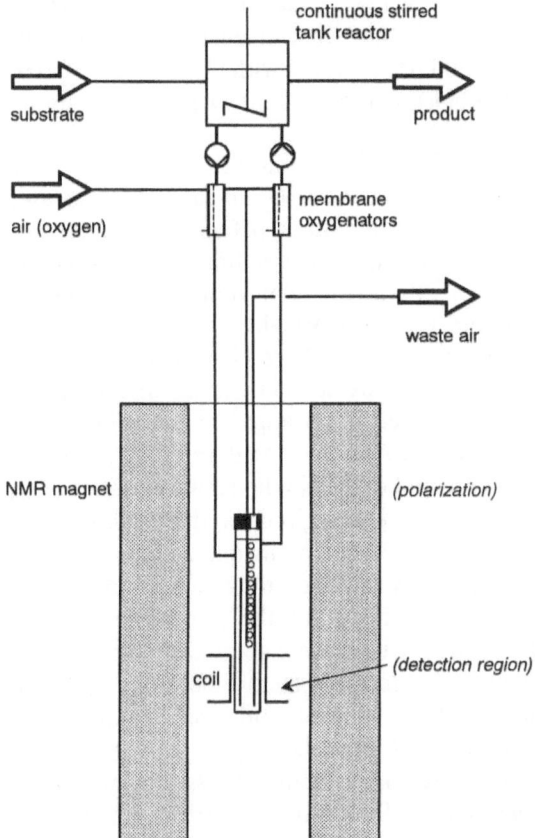

Fig. 3. Principle of a suspended cell reactor system with on line NMR

suspensions is a long established – though still developing – technique. Therefore, although they received relatively little attention, NMR methods have also been developed to study dilute cell suspensions. A near-at-hand solution for NMR studies of dilute cell suspensions was described by Chen and Bailey [30] (Fig. 3). With this setup, a sample stream from a laboratory bioreactor was circulated to the NMR sample chamber in a gas exchange system which permitted maintenance of aerobic conditions for cell densities greater than $30 \, g \, l^{-1}$ dry weight consuming oxygen at a specific rate of $5 \, mmol \, g^{-1} \, h^{-1}$. Sensitivity, however, was not optimized. ^{31}P NMR spectra with acceptable SNR at routine cell densities of $25 \, g \, l^{-1}$ dry weight were obtained in about 10 min. The drawback of this system is that the cells are uncontrolled in the by-pass for about 1.5 min, whence significant changes of metabolism may occur.

2.5 Suspended Cell In Situ Reactor Systems for NMR-Studies on Microorganisms

To avoid undefined conditions of cells in a bypass through the NMR magnet, in situ reactors seem preferable. As early as 1981, Balaban et al. [31] reported a homebuilt NMR-probe for use in situ in the NMR magnet, in which suspensions (10 ml) of $36 \, g \, l^{-1}$ dry weight could be measured with ^{31}P NMR in 10 min. Mitochondria and renal cortical tubules could be kept in a viable state for several hours.

Meehan et al. [32] developed a flask-shaped continuous cell cultivator for operation in a horizontal bore NMR magnet. This 300 ml cultivator was used as a true in situ bioreactor equipped with temperature, pH and volume control. It was suited for the prolonged aerobic continuous cultivation of *Saccharomyces cerevisiae* at a reported cell density of 7% intracellular volume. Assuming the dry weight to represent 30% of the total cell weight, this corresponds to a density of about $23 \, g \, l^{-1}$ dry weight. The NMR receiver coil was located within the cultivator, the contents of which were mixed by an air-driven turbine coupled to an impellor. The mixing resulted in an apparent reduction in the spin-lattice relaxation time T_1, leading to an increased sensitivity. This is because nuclear spins in the detection region of the NMR coil were continuously replaced by fully polarized spins from other regions in the reactor. Thus ^{31}P NMR spectra of reasonable quality could be obtained in as little as 8 min.

A similar continuous flow bioreactor system was simultaneously developed by De Graaf et al. [33]. However, in contrast to the previous study, the dimensions and flow rates used with this reactor were specifically designed for optimal NMR sensitivity. The reactor fits into the 89 mm bore of a standard vertical wide-bore superconducting NMR magnet and was operated with a slightly modified 20 mm standard NMR probehead (Fig. 4). The all-glass reactor of approx. 360 ml volume consisted of a cylindrical vessel (length 14 cm, diameter 7 cm) in connection with a 20 mm NMR tube of 5.5 cm length attached to the vessel bottom along the vertical long axis of the vessel. The 20 mm tube

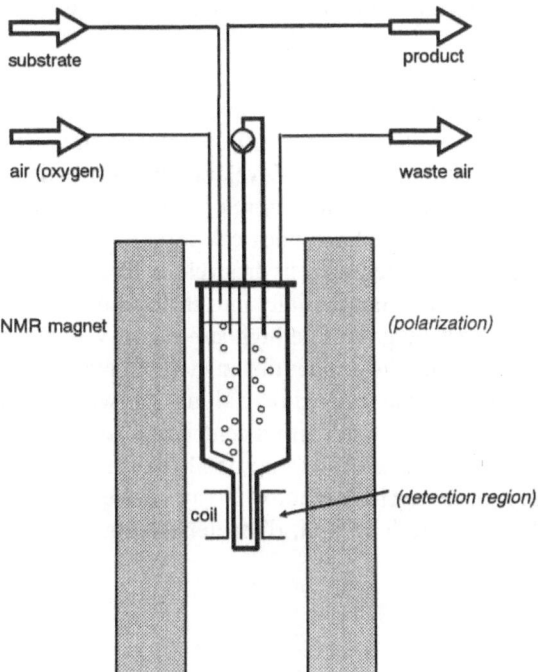

Fig. 4. Principle of a suspended cell in situ NMR reactor system

fitted into the sample chamber of the NMR probehead. A concentric 10 mm tube extended over the whole length of the vessel and 5 cm into the 20 mm tube. A small centrifugal pump with its inlet submersed in the vessel and its outlet connected to the upper end of the 10 mm tube was used to circulate the fermenter contents with a rate of 150 to 600 l h^{-1}. Since the stainless-steel lid on top of the reactor contained ports for measuring and control devices (pH electrode, titration tubing etc.) the reactor could be fully operated in situ in the NMR magnet. A very high sensitivity was achieved by using the reactor vessel as a prepolarizing volume. The dimensions and flow rates could be chosen such that the mean residence time of nuclear spins with T_1 values of 440 ms up to 1.8 s was long enough to ensure complete polarization. At the same time, the mean residence time in the measuring chamber caused T_2^* of the NMR signal to be adjustable between 18 and 72 ms since the volume of the detection region of the coil was about 9 ml. These values are ideal for ^{31}P NMR studies in vivo since typical T_1 values are 400–1000 ms and typical intrinsic T_2^* values are less than 20 ms. The high flow rate enabled one to apply r.f. pulses at a very high rate of once every 72 ms. Thus, from an anaerobic cell suspension with density only 2 g l^{-1} dry weight, ^{31}P spectra with acceptable SNR could be obtained in only 30 min. The spectra showed significant differences with spectra obtained from dense cell suspensions.

2.6 Suspended Cell In Situ Reactor System for NMR-Studies at High Cell Densities

Based on the development of the continuous flow bioreactor system designed for optimal NMR sensitivity by de Graaf et al. [33], an in situ NMR reactor system for continuous cultivation of high cell densities was developed by Hartbrich et al. [34, 35]. This all-glass reactor fits into the 89 mm bore of a standard vertical wide-bore superconducting NMR magnet. The cylindrical part of the reactor (length 250 mm, diameter 70 mm) is operated as a hydrocyclone (see Fig. 5).

The microbial cell suspension is pumped tangentially into the cyclone reactor by a recycling pump outside the NMR magnet. The high superficial velocity of the cell suspension of up to $5 \, \mathrm{m \, s^{-1}}$ in the tangential inlet (diameter 9 mm) produces a turbulent primary whirl which runs down along the wall of the cyclone reactor. This results in a centrifugal field with a centrifugal acceleration of up to 25 g [35]. That is why the gas phase of the microbial cell suspension is forced inwards to the vertical axis of the cyclone reactor. To avoid the transport of gas bubbles into the NMR tube (length 55 mm, diameter 20 mm) which is

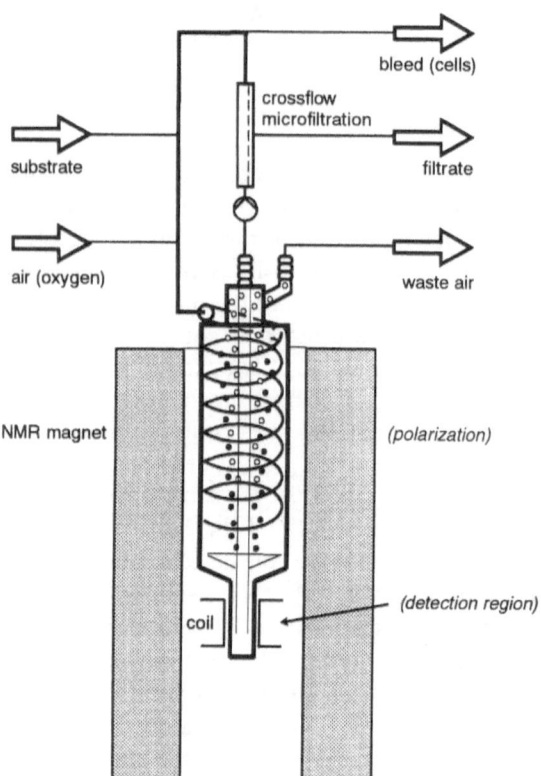

Fig. 5. Principle of a membrane-cyclone-reactor system for in vivo NMR studies at high cell densities

attached to the cyclone reactor bottom along the vertical axis, a conical plate with a gap of 2 mm to the wall of the cylindrical part is fixed at the bottom of the cyclone reactor. The conical plate reverses the inner part of the whirl which mainly consists of the gas phase. The separated gas is withdrawn at the top of the cyclone reactor. The cell suspension without gas bubbles is pumped through the NMR tube. The NMR tube is connected with a concentric tube extended over the whole length of the cyclone reactor and 50 mm into the NMR tube. The cell suspension is then recycled to the tangential inlet by the recycling pump. To achieve cell accumulation in this continuous reactor system, a crossflow micro-filtration unit is installed in the recycling tube outside the NMR magnet. The unit (pore size 0.45 μm, area 0.14 m^2) consists of a ceramic tube module with 19 channels each of which has an inner diameter of 2.7 mm. Measuring and control devices of the membrane-cyclone-reactor (pH-sensor and titration supply, pO_2-sensor and two-phase nozzle, T-sensor and heat exchanger, substrate supply, bleed removal and filtrate removal) are integrated into the recycling tube. For level control of the membrane-cyclone-reactor a potentiometric electrode is used in the cyclone-reactor. The cyclone-reactor with a liquid volume of 900 ml acts as a prepolarizing volume to achieve a high NMR sensitivity, since the dimensions and circulation flow rates (100–1240 $l h^{-1}$) are chosen such that the mean residence time of nuclear spins with T_1 values of up to 1000 ms is long enough to ensure complete polarization. At the same time the mean residence time in the NMR tube (measuring chamber) is adjustable between 26 and 324 ms, since the volume of the detection coil is about 9 ml. For in vivo ^{31}P NMR studies an optimal circulation flow rate of 640 $l h^{-1}$ is adjusted which enables one to apply r.f. pulses at a very high rate of once every 50 ms. The mean residence time of the cell suspension in the circulation tube is 2.3 s at a circulation flow rate of 640 $l h^{-1}$. Although the measurement of the residence time distribution of the membrane-cyclone-reactor demonstrates an ideally mixed behaviour with respect to typical residence times at a circulation flow rate of 500 $l h^{-1}$ [36], concentration gradients in the circulation tube could appear at high cell concentrations and at high cell specific uptake activities if the circulation flow rate decreases. For in vivo ^{13}C NMR studies the mean residence time in the NMR measuring chamber must be set to 200 ms (flow rate 160 $l h^{-1}$), because the typical intrinsic T_2^* values are much higher than for in vivo ^{31}P NMR studies. To avoid concentration gradients in the circulation tube at low flow rates of about 160 $l h^{-1}$ the flow rates in the NMR-tube and the circulation tube must be decoupled. For in vivo ^{13}C NMR studies the cyclone reactor was therefore changed: the conical plate at the bottom of the reactor was lifted 60 mm and the concentric tube extended over the whole length of the cyclone reactor was perforated with ten holes (diameter 5 mm) below the conical plate. This results in the desired optimal flow rate of about 160 $l h^{-1}$ in the NMR tube and about 600 $l h^{-1}$ in the circulation tube.

Decoupling the residence times of cells and medium in a membrane reactor, which means the ability to select independently the residence times of medium (via control of substrate supply) and cells (via control of bleed removal), offers the chance to study cells at nearly every desired metabolic state in a steady state, even at low growth rates. The cell density in the membrane reactor, which is controlled by the ratio K of the residence time of the cells to the residence time of the medium ($K \geq 1$) and the cell yield, is limited by two factors:

a) the accumulation of non-metabolizing cells and cell debris with extended residence times of the cells, and
b) the maximum oxygen transfer rate attainable in the reactor.

The difference between metabolic activity of cells with and without cell retention in a continuous total mixed system can be determined by comparing steady state results in a membrane reactor with a simulation of steady state results based on macrokinetic data from chemostat experiments. For the glucose conversion of the anaerobic ethanol producing bacterium *Zymomonas mobilis* the mass balances of the membrane reactor were solved using the macrokinetic data of Weuster-Botz [37] and compared to the steady state results in the membrane reactor [38]. It was found that for *Zymomonas mobilis* a residence time ratio K of more than 20 results in noticeable differences in the steady state cell concentration due to changes in metabolic activity (see Fig. 6). Therefore experiments are performed at $K \leq 10$. The maximum achievable cell concentration of $23 \, g l^{-1}$ *Zymomonas mobilis* dry weight at $K = 10$ reduces the acquisition time by a factor of 100 compared to a continuous suspended cell cultivation without

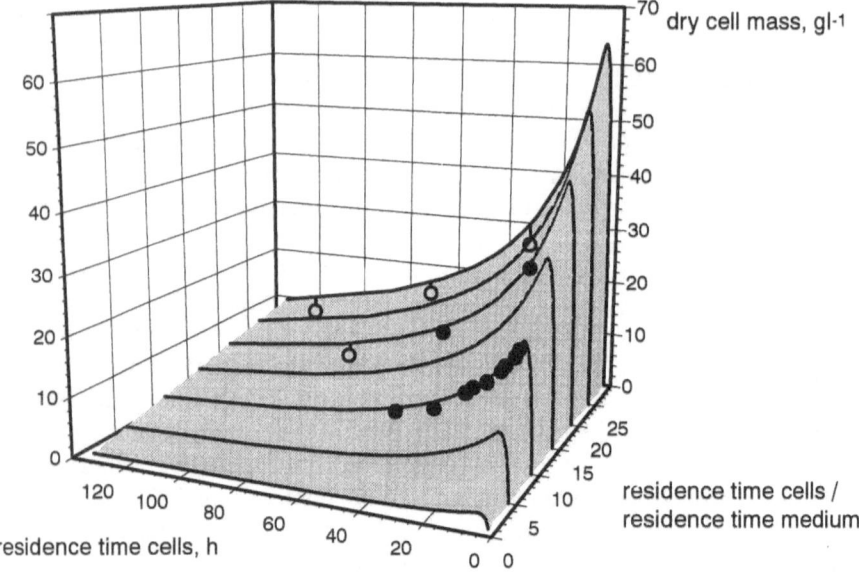

Fig. 6. Reaction engineering analysis of *Zymomonas mobilis* in a total mixed membrane reactor system (simulation based on the macrokinetic data from chemostate experiments [4], ● steady state concentration with a deviation $< 5\%$, ○ steady state concentrations with deviations $> 5\%$)

cell retention at the same reaction conditions (the acquisition time α (cell density)$^{-2}$).

The second limitation of the maximum cell density of aerobic microorganisms in the reactor is the oxygen transfer rate. The cyclone reactor is designed for high oxygen transfer rates by maximizing the gas-liquid surface area a and the transport coefficient k_L in the turbulent primary whirl. Additionally the pressure in the centrifugal field is increased. This increases the driving oxygen concentration difference between gas-liquid surface and liquid bulk phase.

The volumetric oxygen transport coefficient $k_L a$ of the cyclone reactor for in vivo NMR studies has been determined as a function of specific energy input and gas input [39]. At a circulation flow rate in the membrane-cyclone reactor of 600 l h^{-1}, a $k_L a$ and an oxygen transfer rate of 0.55 s^{-1} and 15 g O$_2$ l^{-1} h^{-1}, respectively, is achieved at a gas flow rate of 300 l h^{-1} air (energy input: 25 kW m^{-3}). For example this oxygen transfer rate permits an adequate oxygen supply of about 50 g l^{-1} *Corynebacterium glutamicum* dry weight (maximum of cell specific oxygen demand: 300 mg O$_2$ g^{-1} h^{-1}).

The study of phosphorus limitation in *Zymomonas mobilis* is an example of in vivo ^{31}P NMR spectroscopy using the membrane-cyclone-reactor configuration [40]. The good SNR of the ^{31}P spectra (see Fig. 7) enabled the quantitative determination of the intracellular concentrations of at least two sugar-phosphates (SP) of the Entner-Doudoroff-pathway, phosphate (and extracellular phosphate), nucleosidediphosphates (NDP) and nucleosidetriphosphates (NTP), uridinediphosphate-sugars (UDP-sugar), the sum of NAD and NAD(H) and a cyclic pyrophosphate. Using this membrane-cyclone-reactor configuration, cyclic pyrophosphate in *Zymomonas mobilis* was successfully detected for the first time. Increasing the mean residence time of the cells in the membrane-cyclone-reactor from 5.8 h to 16.0 h without varying the mean residence time of the medium (1.9 h) resulted in a drastic reduction of the extracellular phosphate concentration from 0.5 mmol l^{-1} to 0.003 mmol l^{-1} in about 8 h (see Fig. 7). The steady state intracellular phosphate concentration of 1.1 mmol l^{-1} was, however, not changed. The dynamic response of intracellular metabolite concentrations to this change in the residence time of the cells was recorded using the ^{31}P signals. Changes in the intracellular concentrations were detected for sugar-phosphates, UDP-sugars and cyclic pyrophosphate (see Fig. 7).

The study of the dynamics of the conversion of ^{13}C labelled glucose by *Corynebacterium glutamicum* is an example of in vivo ^{13}C NMR studies using the membrane-cyclone-reactor configuration [41–43]. At defined steady state conditions (mean residence time medium 20 h, mean residence time *Corynebacterium glutamicum* cells 100 h, cell mass 25 g l^{-1} dry weight, glucose 0.9 g l^{-1}, L-lysine 18 g l^{-1}) the substrate supply was changed from unlabelled glucose to a substrate with labelled glucose (labelling in position 6, 10% of the glucose in the substrate supply). After 90 min the substrate supply was changed to unlabelled glucose. The dynamics of labelling of the different C-atoms in L-lactate, L-glutamate, succinate and L-lysine were recorded using the in vivo ^{13}C-signals (see Fig. 8). It is currently being attempted to calculate, using an appropriate

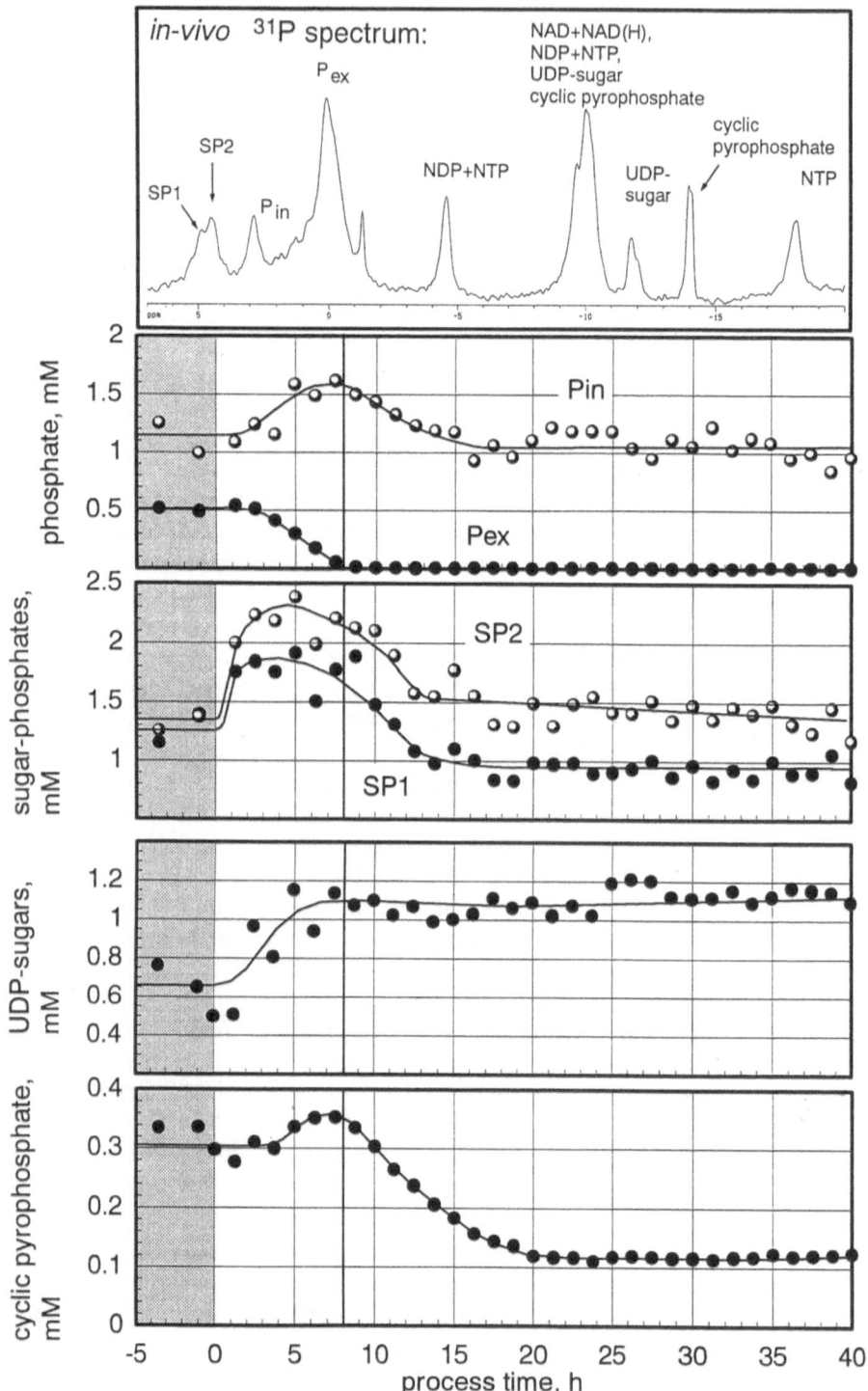

metabolic model, the distribution of the C-fluxes in the biosynthetic pathways to L-lysine (succinyl-diaminopimelate pathway and diaminopimelate-dehydrogenase pathway) on the basis of the measured differences in the labelling rates of the C-atoms of L-lysine.

2.7 Comparison of Reactors for In Vivo NMR

Reaction engineering data for different reactor systems applicable for in vivo NMR studies are summarized in Table 1. The main features of reactor systems for in vivo NMR application are

- high cell densities to achieve high metabolite concentrations in the NMR,
- high mass transport rates to avoid concentration gradients in the reactor system due to mass transport limitations, and
- the application of a prepolarizing volume in the NMR magnet together with a controlled flow rate in the measuring chamber for optimizing the signal sensitivity.

That is why the cell density, the oxygen transport coefficient k_La and the r.f. pulse rate are chosen as characteristic parameters for comparison of reactor systems for in vivo NMR.

The acquisition time is not considered as a quantitative parameter, because the acquisition time is fixed intuitively by the authors to achieve a 'SNR with reasonable quality'.

Optimal systems for the long-term cultivation of mammalian cells in situ in the NMR magnet at high cell densities are hollow fiber bioreactor systems. The maximum k_La of $0.056 \, s^{-1}$ (calculated indirectly based on the data given by Drury et al. [44]) is high compared to other membrane reactor configurations (up to $0.007 \, s^{-1}$) [45]. Circulation of cells through the detection region of the NMR coil had not been realized in hollow fiber bioreactor systems up to now. This results in low r.f. pulse rates of $1 \, s^{-1}$ and acquisition times of about 60 s to achieve an acceptable SNR. Considering the slow metabolic conversion rates of mammalian cells compared to microbiol cells, slow transitional phase studies could well be performed with these systems.

The application of hollow fiber bioreactor system for in vivo NMR studies of microbial cells is not possible due to the higher metabolic activity of microorganisms, which results in severe mass transport limitations. Cell densities of

---

Fig. 7. In vivo ^{31}P NMR spectrum at process time 0 and the dynamic answer of intra- and extracellular phosphate and intracellular sugar phosphates (SP1, SP2), UDP-sugars and cyclic pyrophosphate to a change in mean residence time to the cells from 5.8 h to 16 h (at process time 0) using an in situ membrane-cyclone-reactor configuration (transitional phase study). Reaction conditions: temperature 30 °C, pH 5.5, glucose in the substrate $130 \, g \, l^{-1}$, first steady state: residence time medium 1.9 h, cell mass $4.1 \, g \, l^{-1}$ dry weight, glucose $80 \, g \, l^{-1}$, ethanol $28 \, g \, l^{-1}$; second steady state: residence time medium 1.9 h, cell mass $11.8 \, g \, l^{-1}$ dry weight, glucose $30.7 \, g \, l^{-1}$, ethanol $65.6 \, g \, l^{-1}$

in-vivo 13C spectrum:
L-Lysine C2
L-Lysine C6
succinate C2, C3
L-glutamate C4
L-Lysine C3
L-glutamate C3
L-Lysine C5
L-Lysine C4
L-lactate C3
L-alanine C3
ppm 55 50 45 40 35 30 25 20

glucose
PEP
pyruvate
oxalacetate
2-oxo-glutarate
succinate
L-aspartate
L-aspartate-semialdehyde
pyruvate
L-2,3-dihydropicolinate
L-2-amino-6-ketopimelate
D,L-diaminopimelate
L-lysine
13C-glucose
substrate supply
C6
L-lactate
peak area
C3
C2
C1
L-glutamate
C4
C3
C2
C1
C5
succinate
C2,
C3
L-lysine
C5
C3
C2
C4
C6
C1
0 3 6
process time, h

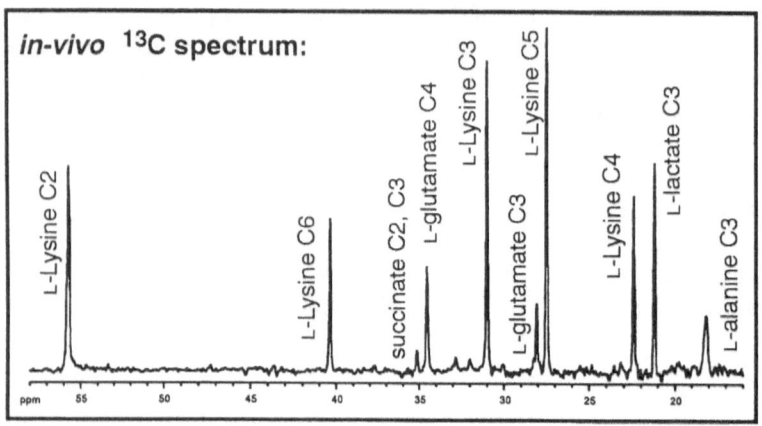

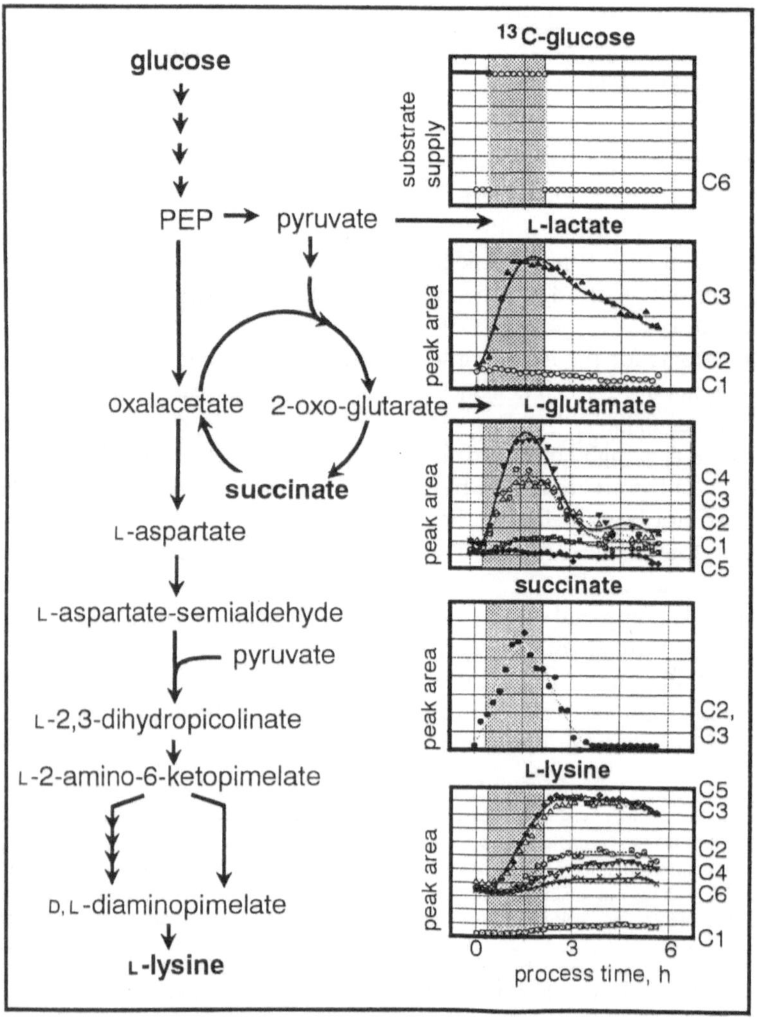

up to 27 g l^{-1} dry weight are reported for suspended cell in situ reactor systems and suspended cell systems with on line NMR. The suspended cell in situ reactor systems have the advantage of achieving higher r.f. pulse rates than a reactor with on line NMR (up to factor of 10). This is why acceptable aquisition times are possible even at very low cell densities (30 min at 2 g l^{-1} dry weight [33]). At high cell densities, aquisition times of about 8 min can be applied. This enables slow transition phase studies. Data for the oxygen transfer rate of the reported reactors are not given, but it could be assumed that the maxima of the oxygen transport coefficients are comparable to stirred tank reactors ($k_L a$ up to 0.2 s^{-1}).

High cell densities (up to 50 g l^{-1} dry weight), high oxygen transport coefficients ($k_L a$ up to 0.56 s^{-1}) and high r.f. pulse rates (up to 20 s^{-1}) can be achieved using the membrane-cyclone-reactor configuration as in situ suspended cell system for in vivo NMR studies. Another feature of this reactor system is the ability to control the mean residence time of the cells (= (growth rate μ)$^{-1}$) independently of the residence time of the medium. This offers the chance to study microbial cells at nearly every desired metabolic state in a steady state or at controlled transitional phases and makes the membrane-cyclone-reactor configuration a near-perfect reactor system for in vivo NMR studies of microbial cells.

Even though the reactor systems for in vivo NMR studies of mammalian cells as well as microbial cells have been optimized in the last few years, the number of in vivo identified intracellular metabolites is still limited to the major pools. In the case of ^{31}P NMR the concentrations of phosphate, ATP, cyclic pyrophosphate and sum parameters for, for example, phosphoesters, sugar phosphates, NTP, NDP, NAD and NAD(H), phospholipids and UDP sugars can be measured in vivo. In the case of ^{13}C NMR only the labelled C-atoms of metabolites which exceed intracellular concentrations of about 0.1 to 1 mmol l^{-1} (depending on the concentration of labelled substrate) can be identified.

In contrast to in vitro NMR studies using cell extracts, the intracellular pH of metabolizing cells is controlled by the cells and is not freely selectable. Therefore non-optimal pH values from the viewpoint of signal separation of titratable compounds may be present. For example, this is the case for *Zymomonas mobilis* where intracellular pH does not rise above 6.5 whereas pH > 8.0 is required for optimal separation of sugar phosphate signals [46].

In vivo NMR studies have, however, already provided a wealth of information on metabolic events in living organisms and many biotechnologically relevant studies using continuously improving instrumentation will be expected in the coming years.

Fig. 8. ^{13}C in vivo NMR study of the dynamics of the conversion of ^{13}C labelled glucose (labelled in position 6) by *Corynebacterium glutamicum* using the membrane-cyclone-reactor configuration (steady state conditions: mean residence time medium 20 h, mean residence time cells 100 h, cell mass 25 g l^{-1} dry weight, glucose 0.9 g l^{-1}, L-lysine 18 g l^{-1}). The labelled C-atom is indicated by Cl, C2 etc

Table 1. Comparison of reactors for in vivo NMR studies

In vivo NMR system	Operation mode	Organisms	Cell density $g\,l^{-1}$dry wt	k_La s^{-1}	r.f. Pulse rate s^{-1}	Identified metabolites	References
Air lift system	batch	*E. coli*	18	0.008	—	—	[19]
Hollow fiber system	continuous with total cell retention	EAT cells	$20\,10^7$ cells ml^{-1}	0.056	—	ATP phosphate	[25]
		mammalian cells	$4\,10^8$ cells ml^{-1}	—	1	ATP phosphate phosphomonoesters phosphodiesters phosphocreatine diphosphodiesters	[26]
On line system	continuous	*E. coli*	27	—	1.4	sugar phosphate phosphate NDP, NTP NAD (H) UDPG	[30]
Suspended cell in situ reactor system	continuous	*Saccharomyces cerevisiae*	23	—	10	2',3': cyclic nucleosides phosphomonoester phosphate phospholipid ATP	[32]
Suspended cell in situ reactor system	continuous	*Zymomonas mobilis*	2	—	14	sugar phosphates phosphate NDP, NTP NAD(H) UDP sugars	[33]

Membrane cyclone reactor system	continuous with controlled cell retention	*Zymomonas mobilis*	29	—	20	2 sugar phosphates phosphate NDP NTP NAD(H) UDP sugars cyclic pyrophosphate	[40]
	continuous with controlled cell retention	*Corynebacterium glutamicum*	50	0.56	5	^{13}C labels in L-lactate L-glutamate succinate L-lysine L-alanine	[41]

3 Reaction Engineering of Rapid Sampling
for Intracellular Metabolite Analysis

3.1 Principles of Rapid Sampling

For the characterisation of metabolic processes in microorganisms, reliable information about the concentrations of intracellular metabolites is of crucial interest. The principal steps of a sampling procedure for intracellular metabolite analysis are: obtain a representative sample; inactivate the metabolism; extract the metabolites; denature and/or separate the enzymes of the cells; begin the analytical procedure (see Fig. 9).

The application of (invasive) sampling techniques to measure reliable intracellular metabolite concentrations can only be successful if

1. a representative sample can be taken from a controlled reactor without disturbing the reaction,
2. the inactivation of the metabolism in the sampled cells is rapid as compared to the metabolic reaction rates to avoid uncontrolled reactions in the sampling device,
3. the extraction of the intracellular metabolites and the denaturation and separation of the intracellular enzymes is complete,

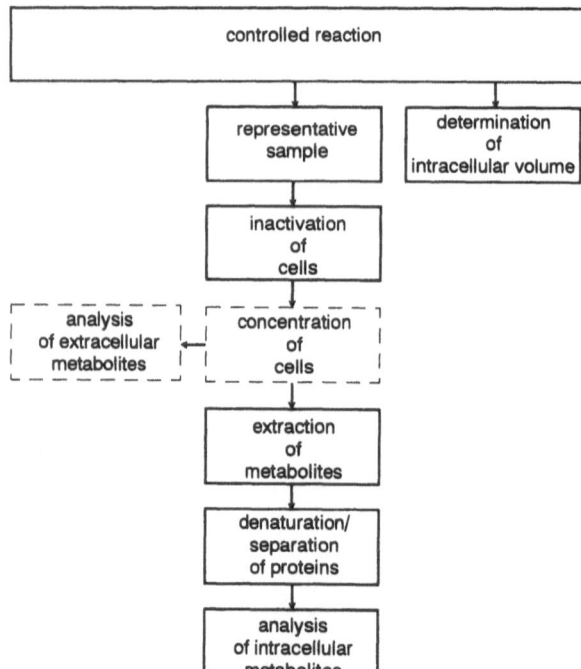

Fig. 9. The principal steps of a sampling procedure for intracellular metabolite analysis

4. the sampling, inactivation and extraction procedure does not affect the stability of the metabolites, and
5. the dilution by inactivation and/or extraction is controlled, reproducible and minimal.

If rapid dynamic metabolic reactions are under study, the sampling rate must be made high enough.

The requirements for the application of rapid sampling systems for reliable intracellular metabolite analysis are high. It must be pointed out that many metabolic reactions, especially catabolic reactions and reactions which are involved in the energy metabolism, have high turnover rates: cytosolic glucose is converted at approximately $1 \, mmol \, l^{-1} s^{-1}$, cytosolic ATP at approximately $1.5 \, mmol \, l^{-1} s^{-1}$ [47, 48]. Considering the reported intracellular levels of glycolytic metabolite concentrations and cytosolic ATP concentrations of up to $4 \, mmol \, l^{-1}$ [49, 50], an inactivation time far below 1 s is necessary. Although the primary handling of cells in the sampling device may not induce significant changes of metabolites within parts of a second, since the extracellular reaction conditions may not be changed very drastically in comparison to the reaction conditions in the reactor, it is important to be aware of this problem.

That is why the classical approach of rapid inactivation within a few milliseconds has become the direct quenching of an incubation mixture or a sample in liquid nitrogen or liquid CO_2; the heat transfer rate is maximized by a high driving force (temperature difference) and by a high sample-liquid surface area (spraying of the sample) [51].

The high turnover rates of catabolic reactions require not only a minimization of the inactivation time, but also a high sampling rate if fast dynamic metabolic reactions have to be measured. The maximum sampling rate is limited by the mixing time, which can be achieved in the reactor or the incubation device to fulfill the requirement for a representative sample.

Considering the six principal steps of a sampling procedure for intracellular metabolite analysis (see Fig. 9), it is important to minimize the probability of errors of each step, because the error propagation is multiplicative. For example, a mean relative error of 1.5% for each step results in a total error of 11.4% after six steps.

For analytical determination of intracellular metabolites, mainly HPLC-techniques or capillary electrophoresis (analysis of nucleotides) and enzymatic determinations are applied [45–53]. The precision of these analytical determinations is affected by the problem of low metabolite concentrations in the sample. In most cases the level of intracellular metabolite concentrations is lower than $1 \, mmol \, l^{-1}$. In addition the intracellular volume is only about 1–10% of total volume of the sample, thus reducing the metabolite concentrations again, especially if continuous cultivation techniques are applied.

Concentration steps (for example evaporation under vacuum) or amplification reactions to improve the sensitivity of the analytical determinations (for example 'enzymatic-cycling') must usually be applied [47, 54]. In general this increases the estimation error.

A favorable solution to the problem of low metabolite concentrations is the introduction of a separation step (microfiltration or centrifugation) into the sampling procedure after inactivation of the metabolism of the cells (see Fig. 9). The main problem is to perform this separation step in such a way (preferably at temperatures below $-30°$ C) that the inactivated status of the cells is preserved and the cells are not destroyed [47, 49]. An additional advantage of applying such a separation step is the possibility of distinguishing between intracellular and extracellular metabolite concentrations.

Research work on sampling systems has been focused up to now on the development of systems for process monitoring and control [55, 56]. Such systems must, first of all, be safe to avoid any contamination risk and, secondly, be easy to handle in order to obtain representative, reliable samples in a reproducible manner. The parameters usually measured are substrate(s), cell mass and product(s) in the medium by manual offline analytical determination or by online analysis systems. Sampling systems for monitoring and control have been reviewed recently [55, 56]. These sampling systems designed for process monitoring and control cannot be applied to the determination of intracellular metabolites, mainly because of an insufficient inactivation time.

The few published sampling systems that are adequate for intracellular metabolite analysis will be briefly reviewed and discussed below.

3.2 Rapid Sampling by Methanol Spraying

De Koning and van Dam [47] developed a method for the determination of changes of glycolytic metabolites in yeast on a subsecond time scale.

Incubations with glucose for less than 10 s were performed with a freeze-quench apparatus with a four-jet tangential mixer. The apparatus was equipped with two stainless steel syringes and worked with a flow rate of 1 ml s^{-1} after mixing. The internal volume of the incubation tubing was varied to change the incubation time from 0.015 to 5 s. This results in one sample for each experiment.

Glucose metabolism was stopped within 100 ms by spraying 10–15 ml samples of incubated yeast suspension in a stirred solution of 60% methanol kept at $-40°$ C. After quenching, the cells were separated by centrifugation at $-20°$ C and the drained pellet was suspended in 2.5 ml of 100% methanol at $-40°$ C.

As an alternative to centrifugation, a filtration at $-40°$ C was used. After washing the separated cells three times with 5 ml of cold 60% methanol the filter was transferred to precooled 100% methanol, the cells were removed from the filter and proteins were denatured. To inhibit Mg^{2+}-dependent partly chloroform-resistant enzyme activities, 20 µl of 200 mmol l^{-1} EDTA (pH 7) was added. For denaturation of the proteins, 1 ml of precooled chloroform was used and the sample tubes were frozen and stored at $-70°$ C.

The extraction of metabolites was subsequently performed using chloroform at $-40°$ C. To achieve complete permeabilization and extraction the sample tubes with the added chloroform were shaken for 45 min at $-35°$ C. The cells were pelleted in the chloroform phase by centrifugation. The supernatant was collected and was extracted once with 15 ml diethyl ether to remove traces of lipids. The volume of the chloroform sample was reduced under vacuum, starting with the frozen sample to prevent excessive boiling due to the remaining ether. Shortly before the determination of metabolites, the sample was centrifuged in order to remove possible small precipitates. Enzymatic determinations of metabolites were performed in 50 mmol l^{-1} triethanolamine, pH 7.6, following the changes in NAD(P) H at 340 nm as described by Bergmeryer [57]. The analyses were performed on an autoanlyzer system, thus resulting in higher reproducibility. The routine detection level was 1 μmol l^{-1} in the sample, and the differences between duplicates were less than 3%. Variation between different extractions of the same incubation was usually within 10%.

The metabolites under consideration were glucose-6-phosphate (G6P), fructose-6-phosphate (F6P), dihydroxyaceton-phosphate (DHAP), fructose-1, 6-bisphosphate (FBP) pyruvate, phosphoenolpyruvate (PEP), 2-phosphoglycerate (2PGA), 3-phosphoglycerate (3PGA), AMP, ADP, ATP, NAD and NADH. The intracellular concentration of PEP was below the detection level, 2PGA and 3PGA were at the detection level, highest intracellular concentrations of catabolic metabolites were measured for glucose (0.9 mmol l^{-1} after 0.8 s of incubation) and G6P (0.3 mmol l^{-1} after 0.8 s of incubation).

Seiler et al. [49] used methanol spraying for the determination of intracellular metabolites in *Escherichia coli*. In contrast to de Koning and van Dam [47] they cultivated the cells in a controlled stirred tank reactor before quenching the samples in $-40°$ C methanol. The sampling device is not described and data about the inactivation times are not given. The samples were extracted using perchloric acid extraction, because the chloroform extraction was shown not to be feasible for *E. coli*.

The metabolites under consideration were glucose-6-phosphate (G6P), fructose-1,6-bisphosphate (FBP), pyruvate, phosphoenolpyruvate (PEP), ADP and ATP. The detection level of metabolic intermediates determined by NAD(P) H based enzymatic assays was 1–5 μmol l^{-1} intracellular concentration for most metabolites. It was found that the intracellular concentration of G6P was below the detection level at the stationary phase and reached a concentration of 3.5 mmol l^{-1} at the logarithmic phase of a pH-controlled batch culture. The other intracellular metabolites remained essentially unchanged at a level lower than 0.5 mmol l^{-1} except ATP, for which an intracellular concentration of 3 mmol l^{-1} was measured.

3.3 Harvesting Valve for Rapid Sampling from Bioreactors

In order to minimize the time elapsed between cell harvesting from a controlled reaction in a bioreactor and the inactivation of the metabolism of the cells, Theobald et al. developed a sampling device using a miniature valve [48]. A hypodermic needle was inserted into the reactor and the miniature valve was coupled with a HPLC capillary with an inside diameter of 0.7 mm. The samples were taken with vacuum sealed, precooled glass tubes at $-20°$ C containing 1 ml perchloric acid (35% w/v) and 10–15 glass beads with a diameter of 4 mm to improve the heat transfer capacity. The inactivation time was estimated to be 500 ms. The sampling rate of this manual sampling procedure (setting a new precooled glass tube to the miniature valve, opening the valve, closing the valve, taking the tube away from the miniature valve) was 0.2 s^{-1}.

For optimal extraction, three freeze-thaw cycles ($-20°C/0°C$) were performed. A 2 mol l^{-1} KOH solution with 0.5 mol l^{-1} imidazol was applied for neutralization, and the resulting $KCeO_4$ precipitate was removed by filtration.

This method was used for the determination of adenine nucleotides and catabolic metabolites in yeast cells after glucose injection into a continuously operated stirred tank reactor at a glucose limited steady state (growth rate 0.1 h^{-1}) [48, 58, 59]. Adenine nucleotide concentrations were determined using HPLC. The determination of the cytosolic part of adenine nucleotide concentrations was achieved by the use of precooled $HgCl_2$ ($-20°C$) for inactivation of the cells in the sampling tubes instead of perchloric acid followed by permeabilisation of the outer cell membrane with 2 g l^{-1} poly-L-lysine. For determination of G6P, F6P, FBP, pyruvate, PEP and 6-phosphoglycerate (6PG), NAD(P) H-based enzymatic assays with amplification reactions ('enzymatic cycling') were applied [54].

Using this harvesting valve device for rapid sampling from bioreactors, the dynamic properties of these catabolic metabolites were monitored on a time scale of up to 180 s after glucose injection. The rapid increase in intracellular, G6P, F6P, FBP, pyruvate and adenine nucleotide concentrations after glucose injection could not be monitored due to the sampling rate of 0.2 s^{-1}, but the intracellular concentrations of GAP and PEP changed within a time scale of 60 s, and thus the direct monitoring of the increase of these metabolite concentrations was successful. These dynamic data in the time scale of up to 180 s of the measured intracellular metabolite concentrations were satisfactory for identifying the parameters of a structured dynamic model of the glucose metabolism in *Saccharomyces cerevisiae*.

3.4 Sampling Tube for High Sampling Rate

To overcome the problem of high sampling rates for rapid cell harvesting from a controlled reaction in a bioreactor, a sampling tube device was developed [60]. The basic idea was to perform sampling, inactivation and extraction continuously in a tube connected to a reactor, thus fixing fast dynamic reactions at certain positions in the sampling tube.

A sampling probe with an inlet for continuous sampling at the tip of the probe, an inlet for continuous supply of extraction medium (35% w/v perchloric acid precooled to $-25°C$) on the other side of the probe, and an outlet connected to the sampling tube was installed into a standard connecting pipe of a continuously operated stirred tank reactor (see Fig. 10). The extraction medium was enabled to mix with the sample continuously 3 mm from where the sample entered the tip of the sampling probe. At sample flow rates of about 7–10 ml s^{-1} the inactivation of the cells started within 4–5 ms. Using precooled perchloric acid the time for the overall inactivation of the metabolism is reported to be about 200 ms [61]. The sampling tube was filled with water before starting continuous rapid sampling to provide a constant pressure driven continuous flow of sample and extraction medium into the sampling tube. The pressure in the reactor was about 1.5 bar and the pressure in the perchloric acid receiver was about 1.7 bar to achieve a total flow rate of sample mixed with perchloric acid of about 15 ml s^{-1} and a sample to perchloric acid ratio of about 1. The exact flow rates were determined gravimetrically to calculate the dilution factor of the extraction and to transform the position of a sample in the sampling tube to the sampling time. The sampling tube made of polypropylene had an inside diameter of 8 mm, a length of 100 m and was wound up with a diameter of 0.5 m. Sampling was stopped after 3.5 min by closing the valves at the sampling probe. The sampling tube was disconnected and frozen at − 80 °C. To achieve single samples, the frozen sampling tube was divided into identical parts using a circular saw. The sawn parts of the tube with the frozen samples were put into sample flasks for collecting the thawing sample and were stored at − 80 °C. The sampling rate can be selected by the length of the sawn parts of the

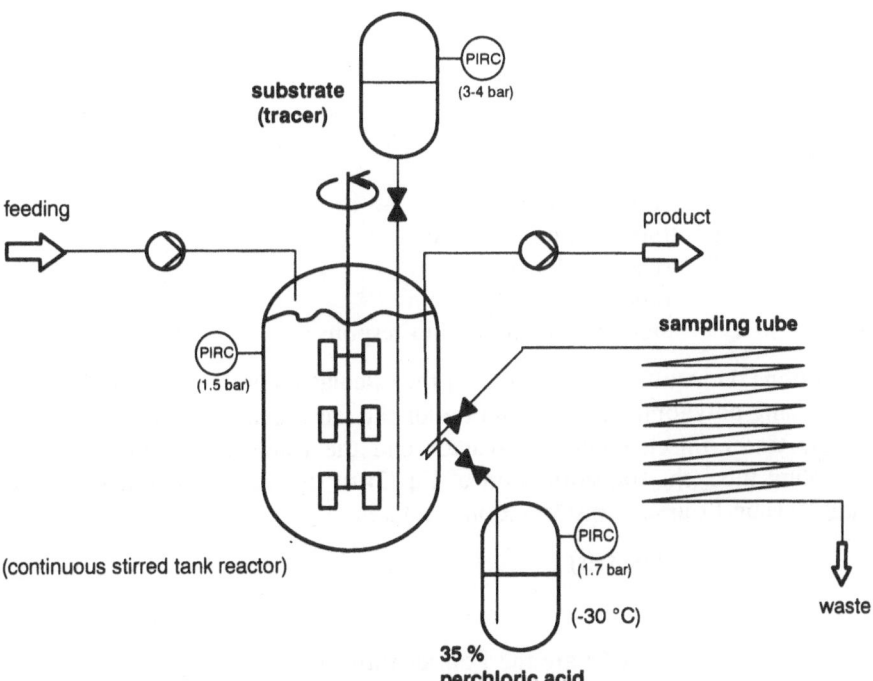

Fig. 10. Principle of the sampling tube device for continuous rapid sampling

tube, for example a length of 0.15 m results in a sampling rate of 2.0 s^{-1}. The limiting factor is the sample size required for the following analysis of the extracted metabolites.

For determination of the concentrations of catabolic metabolites, NAD(P)H-based enzymatic assays with amplification reactions ('enzymatic cycling') were applied after neutralization of the samples with $2 \text{ mol}1^{-1}$ KOH and $0.5 \text{ mol}1^{-1}$ imidazol [54, 57].

The measured concentration-time profiles are distorted due to the axial dispersion in the sampling tube. Therefore a tracer must be injected into the reactor together with the substrate pulse to determine the transfer or weighting function g(t), which describes the residence time distribution of the liquid phase in the sampling tube. For this purpose it is convenient to use the substrate (tracer) pulse itself. This pulse results in a step function, provided the substrate conversion of the microorganisms in the reactor is small compared to the substrate pulse during the complete reaction time to be monitored. This step function is called the tracer input function x(t).

The response function y(t) was calculated from the measured substrate concentration vs position in the sampling tube, using the measured superficial velocity of sample and perchloric acid. The input function x(t) was measured using a conductivity probe at the standard connecting pipe of the reactor instead of the sampling probe at the same reaction conditions (stirrer speed, aeration, injection). Equation (2) describes the correlation between input function x(t) and response function y(t) [62] :

$$y(t) = \int_{0}^{\infty} x(t - t')\, g(t')\, dt \tag{2}$$

where

$x(t)$ tracer input function
$y(t)$ tracer response function
$g(t)$ weighting function
t time
t' time between input and response.

Equation (2) must be deconvoluted for the calculation of the weighting function g(t). Using polynomial approximation for the measured concentration profiles of the tracer injected into the reactor and the tracer measurements in the sampling tube, the transformation of Eq. (2) to the frequency domain (image range) using Fourier transformation results in

$$Y(s) = X(s)\, G(s) \tag{3}$$

where

$X(s)$, $Y(s)$, $G(s)$ are the Fourier transforms of x(t), y(t), g(t).

The now identified weighting function G(s) in the frequency domain can then be used to calculated the function of the intracellular metabolite concentration L(s), applying polynomial approximation for the measured metabolite concentration profile m(t) in the sampling tube and Fourier transformation (l(t) is now the unknown input function and m(t) is the measured response function, see Fig. 11):

$$M(s) = L(s)\ G(s) \tag{4}$$

where

$M(s)$, $L(s)$ are Fourier transforms of $m(t)$, $l(t)$.

The inverse Fourier transformation of L(s) results in the desired function of the intracellular metabolite concentration profile vs time l(t).

An example of applying the sampling tube to monitor the rapid dynamic changes in intracellular metabolite concentrations is the glucose injection into

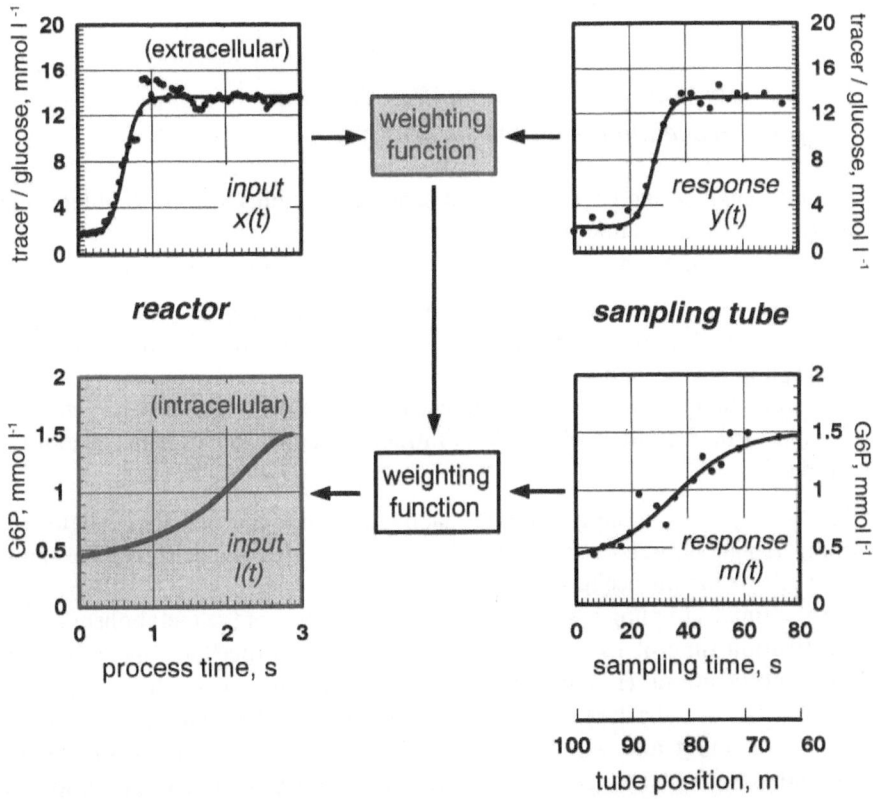

Fig. 11. Determination of the dynamic response of intracellular glucose-6-phosphate concentration to a glucose injection into a glucose-limited steady state cultivation of *Zymomonas mobilis* using the sampling tube. The glucose input function x(t) and the measured glucose concentration profile in the sampling tube (response function y(t)) were used to calculate the transfer or weighting function. This weighting function and the measured G6P concentration profile in the sampling tube were then used to calculate the intracellular G6P concentration profile vs time (● measured concentrations, — polynomial approximation, ▬ calculated profile)

a glucose limited steady state cultivation of *Zymomonas mobilis* bacteria in a continuous stirred tank reactor (see Fig. 11). The change in intracellular G6P concentration in the first 3 s after glucose injection was monitored, taking samples out of the tube for sampling times from 0 to 80 s (position in the sampling tube: 100 to 60 m). The sampling time scale is much larger than the time scale of the intracellular metabolic events due to the dispersion in the sampling tube. The dispersion of the samples in the sampling tube facilitates the monitoring of rapid changes in intracellular metabolite concentrations on a time scale smaller than the time scale which is given by the realized sampling rate. The mixing time of the reactor is no limit for monitoring such fast reactions, because the mixing dynamics of the reactor are considered in the calculation of the weighting function. The drawback of this method for rapid sampling and inactivation is the impossibility of distinguishing between intra- and extra-cellular metabolite concentrations. The detection thresholds of intracellular glycolytic metabolites are about $100 \, \mu mol \, l^{-1}$.

3.5 Comparison of Rapid Sampling Systems for Intracellular Metabolite Analysis

Reaction engineering data of different sampling systems for intracellular metabolite analysis are summarized in Table 2.

The most important characteristics of rapid sampling systems are

- the sampling rate to achieve a high resolution in time,
- the inactivation time of the metabolism to avoid uncontrolled reactions in the sampling device, and
- the possibility of separating inactivated cells before the extraction starts to avoid further dilution of the (low) intracellular metabolite concentration of the cells.

That is why the sampling rate, the inactivation time and the methods for cell separation are chosen as characteristic parameters for comparison of fast sampling systems for intracellular metabolite analysis.

An optimal sampling systems for rapid inactivation of the metabolism of cells (inactivation time of 100 ms) and identification of intracellular metabolites at low concentrations $(1-5 \, \mu mol \, l^{-1})$ is rapid sampling by methanol spraying [47, 49]. The drawback of this sampling technique is the extremely low sampling rate. De Koning and van Dam [47] used incubation tubings with variable internal volume to change the incubation time. They obtained one sample for each short-time experiment. By variation of the incubation time they have successfully monitored dynamics in intracellular metabolite concentrations of *Saccharomyces cerevisiae*. However the reaction conditions are uncontrolled, in contrast to the methods employing stirred tank reactors.

The sampling rate can be improved by applying a harvesting valve device for rapid sampling. The harvesting valve device as developed by Theobald et al.

Table 2. Comparison of rapid sampling systems for intracellular metabolite analysis

Sampling system	Sampling rate s^{-1}	In-activation time ms	Separation of cells	Extraction method	Analysis of metabolites	Identified metabolites	References
Methanol spraying	—	100	filtration at −40°C centrifugation at −20°C	chloroform	enzymatic	G6P F6P DHAP FBP pyruvate AMP ADP ATP NAD(H)	[47]
	—	—	centrifugation at −30°C	perchloric acid	enzymatic	G6P FBP pyruvate PEP ADP ATP	[49]
Harvesting valve	0.2	500	no	perchloric acid	HPLC, enzymatic + amplification	G6P F6P FBP 6PG PEP GAP pyruvate glycogen AMP ADP ATP	[48, 58, 59]
Sampling tube	2	200	no	perchloric acid	enzymatic + amplification	G6P F6P PEP GAP pyruvate	[60]

results in sampling frequencies of 0.2 s^{-1} at an inactivation time of 500 ms [48]. This sampling rate, however, was insufficient for direct monitoring of the rapid increase in several metabolite concentrations in *Saccharomyces cerevisiae* after glucose injection.

Highest sampling rates (2 s^{-1}) at an appropriate inactivation time (200 ms) were reported for the application of the sampling tube for rapid sampling [60]. The rapid increase in intracellular G6P concentration in *Zymomonas mobilis* within 3 s as a response to a step in extracellular glucose concentration was well monitored, thus making this method ideal for monitoring fast reactions with a time scale of seconds or less. The drawback of this method is the impossibility of distinguishing between extra- and intracellular metabolite concentrations.

4 Conclusions and Perspectives

The development of reaction engineering methods in the last few years for in vivo NMR spectroscopy of intracellular metabolites has enabled NMR investigation of microbial metabolism with a time resolution of < 10 min under well defined physiological conditions. This feature facilitates slow transitional ^{31}P-NMR studies. Thus the metabolic effects of transitions in the reaction conditions, for example pH, temperature, substrate concentrations, inhibitor concentrations or growth rate can be monitored directly. This is especially so for the membrane-cyclone-reactor configuration, as the in situ suspended cell system for in vivo NMR measurements offers the chance to study microbial cells effectively in nearly every desired metabolic state, even at very low growth rates.

In the case of in vivo ^{13}C NMR spectroscopy using step functions of labelled substrates or precursors in a balanced steady state of the culture, very interesting studies are expected for evaluating the distribution of C-fluxes in biosynthetic pathways as a function of reaction conditions. Recombinant strains can be characterized on a more detailed level in an in situ reactor system for in vivo NMR spectroscopy under identical defined reaction conditions. A limiting factor, though, for such in vivo ^{13}C-NMR studies will be the high costs of the labelled substrates.

The most important effect on future developments in dynamic modeling is expected to be the development of reaction engineering methods for rapid sampling and inactivation.

Experimental data for the dynamics of intracellular glycolytic metabolite concentrations within seconds after the addition of a substrate pulse to a balanced steady state culture are absolutely necessary to identify the parameters of dynamic models. The sampling tube device for rapid sampling will be a particularly valuable tool because of the possibility of achieving the highest sampling rates (2 s^{-1}) at short inactivation times (200 ms). A reduction of the labour related costs of manual analysis of the samples and an improvement of reproducability is possible if an autoanalyzer system is applied. Future developments of rapid sampling procedures are necessary to achieve a system with high sampling rates (the sampling tube), short inactivation times (the sampling tube or methanol spraying) and a cell separation step (methanol spraying) to distinguish between extra- and intracellular metabolite concentrations.

5 References

1. Gadian D (1983) Nuclear magnetic resonance and its applications to living systems. Clarendon Press, Oxford
2. Srivastava AK, Volesky B (1991) Appl Microbiol Biotechnol 34: 450
3. Shulman RG, Brown TG, Ugurbil K, Ogawa S, Cohen SM, Den Hollander JA (1979) Science 205: 160
4. Den Hollander JA, Brown TR, Ugurbil K, Shulman RG (1979) Proc Natl Acad Sci USA 76: 6096
5. Ugurbil K, Rottenberg H, Glynn P, Shulman RG (1978) Proc Natl Acad Sci USA 75: 2244
6. Harder A, Roels JA (1982) Adv Biochem Eng 21: 55
7. Bloch F, Hansen WW, Packard M (1946) Phys Rev 69: 127
8. Purcell EM, Torrey RV, Pound RV (1946) Phys Rev 69, 37
9. Fribolin H (1988) Ein- und zweidimensionale NMR-Spektroskopie. VCH, Weinheim, Basel, Cambridge, New York
10. Moon RB, Richards JH (1973) J Biol Chem 248: 7276
11. Salhany JM, Yamane T, Shulman RG, Ogawa S (1975) Proc Natl Acad Sci USA 72: 4966
12. Eakin RT, Morgan LO, Gregg CT, Matwiyoff NA (1972) FEBS Lett 28: 259
13. Brown FF, Campbell ID, Kuchel PW, Rabenstein DL (1977) FEBS Lett 82: 12
14. Ernst RR, Anderson WA (1966) Rev Sci Instrum 37: 93
15. Gillies RJ, Ugurbil K, Den Hollander JA, Shulman RG (1981) Proc Natl Acad Sci USA 78: 2125
16. Den Hollander JA, Ugurbil K, Brown TR, Shulman RG (1981) Biochemistry 20: 5871
17. Den Hollander JA, Behar KL, Shulman RG (1981) Proc Natl Acad Sci USA 78: 2693
18. Santos H, Turner DL (1986) J Magn Reson 68: 345
19. Kramer HW, Bailey JE (1991) Biotechnol Bioeng 37: 205
20. Ugurbil K, Guernsey DL, Brown TR, Tobkes N, Edelman IS (1981) Proc Natl Acad Sci USA 78: 4843
21. Gonzalez-Mendez R, Wemmer D, Hahn G, Wade-Jardetzky N, Jardetzky O (1982) Biochim Biophys Acta 720: 274
22. Foxall DL, Cohen JS (1983) J Magn Reson 847: 285
23. Galazzo JL, Bailey JE (1989) Biotechnol Bioeng 33: 1283
24. Fernandez EJ, Mancuso A, Murphy MK, Blanch HW, Clark DS (1990) Ann NY Acad Sci 589: 458
25. Gillies RJ, Scherer PG, Raghunand N, Okerlund LS, Martinez-Zaguilan R, Hesterberg L, Dale BE (1991) Magn Reson Med 18: 181
26. Gillies RJ, Galons J-P, McGovern KA, Scherer PG, Lien Y-H, Job C, Ratcliff R (1993) NMR in Biomed 6: 95
27. Chresand TJ, Gillies RJ, Dale BE (1987) Biotechnol Bioeng 32: 983
28. Briasco CA, Ross DA, Robertson CR (1990) Biotechnol Bioeng 36: 879
29. Briasco CA, Karel SF, Robertson CR (1990) Biotechnol Bioeng 36: 887
30. Chen R, Bailey JE (1993) Biotechnol Bioeng 42: 215
31. Balaban RS, Gadian DG, Radda GK, Wong GG (1981) Anal Biochem 116: 450
32. Meehan AJ, Eskey CJ, Koretsky AP, Domach MM (1992) Biotechnol Bioeng 40: 1359
33. De Graaf AA, Wittig RM, Probst U, Strohhäcker J, Schoberth SM, Sahm H (1992) J Magn Reson 98: 654
34. Hartbrich A, Weuster-Botz D, de Graaf A, Wandrey C (1995) in Kreysa G (ed) Kurzfassungen DECHEMA-Jahrestagungen '95, Band 1, DECHEMA, Frankfurt, pp 116

35. Hartbrich A, Weuster-Botz D, Wandrey C (1995) Verfahren und Anordnung zur Durchführung biotechnologischer Prozesse unter Kultivierung aerober Organismen in hoher Zelldichte, german patent DE P 44.07 440.9–41
36. Bieck T (1994) Ermittlung von Auslegungskennzahlen für einen Zyklon-Bioreaktor, dissertation, Rheinländ-Westfälische-Technische-Hochschule Aachen
37. Weuster-Botz D (1993) Appl Microbiol Biotechnol 39: 679
38. Keiner M (1992) Einsatz und verfahrenstechnische Optimierung eines Membranreaktors zur anaeroben Hochzelldichte-Fermentation, dissertation, Rheinland-Westfälische-Technische-Hochschule Aachen
39. Langöhrig A (1994) Bestimmung des volumetrischen Stoffübergangskoeffizienten $k_L a$ in einem Reaktorsystem zur in vivo NMR Messung, study, Rheinland-Westfälische-Technische-Hochschule Aachen
40. Schmitz G (1995) In vivo ^{31}P NMR Messung der Glucose- und Fructoseverwertung von *Zymomonas mobilis* im Membran-Zyklon-Reaktor, dissertation, Rheinische Friedrich-Wilhelms-Universität Bonn
41. Hartbrich A (1996) Verfahrenstechnische Charakterisierung von Zyklonreaktoren in der Biotechnologie, PhD-thesis, Rheinland-Westfälische-Technische-Hochschule Aachen
42. Arens C (ed) Wissenschaftlicher Ergebnisbericht 1994, Forschungszentrum Jülich, p 414
43. Hartbrich A, Schmitz G, Weuster-Botz D, de Graaf A, Wandrey C (1996) Biotechnol Bioeng submitted for publication
44. Drury DD, Dale BE, Gillies RJ (1988) Biotechnol Bioeng 32: 966
45. Bliem RF, Konopitzky K, Katinger HWD (1991) Adv Biochem Biotechnol 44: 1
46. Strohhäcker J, de Graaf AA, Schoberth SM, Wittig RM, Sahm H (1993) Arch Microbiol 159: 484
47. De Koning W, van Dam K (1992) Anal Biochem 204: 118
48. Theobald U, Mailinger W, Reuss M, Rizzi M (1993) Anal Biochem 214: 31
49. Seiler M, Sauer U, Bailey JE (1994) Intracellular Metabolite Determination in *Escherichia coli* Institute of Biotechnology, ETH, Zürich
50. Ryll T, Wagner R (1991) J Chromatogr 570: 77
51. Chance B, Eisenhardt RH, Gibson QH, Louberg-Holm KK (eds) (1964) Rapid mixing and sampling techniques in biochemistry. Academic press, New York
52. Ng M, Blaschke TF, Arias AA, Zare RN (1992) Anal Chem 64, 15: 1682
53. Skoog K, Hahn-Haegerdal B (1989) Biotech Tech 3, 1: 1
54. Bernovfsky C, Swan M (1973) Anal Biochem 53: 452
55. Mattiasson B, Hakanson H (1993) Trends Biotechnol 11, 4: 136
56. Spohn U, Voss H (1992) in: Präve P, Schlingmann M, Esser K, Thauer R, Wagner F (eds) Jahrbuch Biotechnologie Carl Hanser Verlag München, Wien
57. Bergmeyer HU (1985) Methods of enzymatic analysis. Vol VI and VII, VCH Publishers, Deerfield Beach, Florida
58. Theobald U, Baltes M, Rizzi M, Reuss M (1991) in: Reuss M, Chmiel H, Gilles ED, Knackmus HJ (eds) Biochemical Engineering Stuttgart, Gustav Fischer, Stuttgart, New York
59. Theobald U (1994) Untersuchungen zur Dynamik des Crabtree-Effektes, PhD-thesis, Universität Stuttgart
60. Weuster-Botz D, Wandrey C (1995) Verfahren und Vorrichtung zur Serienprobenahme biologischer Proben, German patent DE: P44 07 439.5-41
61. Kopperschläger G, Augustin HW (1967) Experientia 23: 623
62. Levenspiel O (1993) The Chemical Reactor Omnibook, OSU Book Stores, Inc., Corvallis, Oregon

In Vivo Stationary Flux Analysis by ^{13}C Labeling Experiments

W. Wiechert, A.A. de Graaf

Institute of Biotechnology, Research Center Jülich, D-52425 Jülich, Germany

Advances in Biochemical Engineering
Biotechnology, Vol. 54
Managing Editor: T. Scheper
© Springer-Verlag Berlin Heidelberg 1996

Stationary flux analysis is an invaluable tool for metabolic engineering. In the last years the metabolite balancing technique has become well established in the bioengineering community. On the other hand metabolic tracer experiments using ^{13}C isotopes have long been used for intracellular flux determination. Only recently have both techniques been fully combined to form a considerably more powerful flux analysis method. This paper concentrates on modeling and data analysis for the evaluation of such stationary ^{13}C labeling experiments. After reviewing recent experimental developments, the basic equations for modeling carbon labeling in metabolic systems, i.e. metabolite, carbon label and isotopomer balances, are introduced and discussed in some detail. Then the basics of flux estimation from measured extracellular fluxes combined with carbon labeling data are presented and, finally, this method is illustrated by using an example from *C. glutamicum*. The main emphasis is on the investigation of the extra information that can be obtained with tracer experiments compared with the metabolite balancing technique alone. As a principal result it is shown that the combined flux analysis method can dispense with some rather doubtful assumptions on energy balancing and that the forward and backward flux rates of bidirectional reaction steps can be simultaneously determined in certain situations. Finally, it is demonstrated that the variant of fractional isotopomer measurement is even more powerful than fractional labeling measurement but requires much higher numerical effort to solve the balance equations.

List of Symbols and Abbreviations

A, B, C, D, E, S, P, ...	metabolite names
$A, B, C, D, E, S, P, ...$	absolute molar pool size of metabolites
b_1, b_2	positional fractional carbon labeling of metabolite B with 2 carbon atoms
$b_{00}, b_{01}, b_{10}, b_{11}$	isotopomer fractions of metabolite B with 2 carbon atoms
$v_1^{\rightarrow}, v_1^{\leftarrow}, v_2^{\rightarrow}, v_2^{\leftarrow}$	forward and backward fluxes corresponding to biochemical reaction steps
$\mathbf{x}, \mathbf{x}^{inp}$	vectors of all fractional carbon labels in a metabolic network and all input labels from substrates fed into the system
$\mathbf{X}$	vector of all absolute pool sizes in a metabolic network
$\mathbf{v}^{\rightarrow}, \mathbf{v}^{\leftarrow}$	vectors of all forward and backward fluxes corresponding to metabolic reaction steps
$\mathbf{v}$	overall flux vector comprising $\mathbf{v}^{\rightarrow}$ and $\mathbf{v}^{\leftarrow}$
$\mathbf{v}^{net}, \mathbf{v}^{xch}$	vectors of all net and exchange fluxes corresponding to metabolic reaction steps
$\mathbf{N}$	stoichiometric matrix
$\mathbf{N}^{cnstr}, \mathbf{c}^{cnstr}$	linear constraint matrix and constraint value vector
$\mathbf{P}_i, \mathbf{P}_i^{inp}$	carbon atom transition matrices corresponding to reaction step i
$\mathbf{Q}_i$	bimolecular isotopomer transition tensor corresponding to reaction step i
$\mathbf{I}$	pool size to fractional labeling state mapping matrix
$\mathbf{w}, \mathbf{y}, \mathbf{Y}$	measured fluxes, labels and pool sizes
$\mathbf{M}_w, \mathbf{M}_y, \mathbf{M}_Y$	measurement matrices for fluxes, labels and pool sizes
$\varepsilon_w, \varepsilon_y, \varepsilon_Y$	measurement noise vectors for fluxes, labels and pool sizes

1 Introduction

1.1 Stationary Flux Analysis

The detailed quantitative knowledge of intracellular metabolic fluxes in vivo is of fundamental importance for the study of microbial metabolism and metabolic engineering, which means an engineering approach for the genetic improvement of metabolic processes with respect to desired products [1–3]. In particular the knowledge of stationary intracellular fluxes in vivo is of immediate practical use for

- the verification of enzyme activities and bidirectional reaction steps taking place in vivo,
- the characterization of different physiological states [4–6] in order to achieve an empirical quantitative comparison of regulatory mechanisms,
- the detailed quantitative discrimination between genetically manipulated microorganisms,
- systematic control analysis using methods of metabolic control theory [7–9].

1.2 Two Well-Established Methods

Stationary flux analysis aims at the quantitation of all intracellular fluxes in central metabolism when the microbial system is in a well defined balanced steady state. In recent years the metabolite balancing approach has become popular in the bioprocess engineering community [4, 5, 10, 11]. It is based on direct measurements of the fluxes between the cells and the surrounding medium (henceforth called the *extracellular fluxes*).

On the other hand metabolic tracer experiments have long been used for stationary intracellular flux determination mostly in biochemical research [12–15]. This technique relies on the fractional isotopic enrichment within intracellular metabolites (henceforth called *fractional labeling*) that can be accessed with NMR or mass spectroscopy.

Both methods, the metabolite balancing approach as well as the tracer approach, expose some inadequacies that cannot be overcome with one method alone. While it turned out that some rather unsafe assumptions on energy balancing have to be made for a complete flux analysis based on extracellular fluxes, only relative fluxes can be determined when only labeling data is available. For this reason tracer studies have always been supported by a few directly available flux measurements, but only recently several new developments in reaction engineering have led to a tight integration of both approaches [16–18]. Currently the tracer technique in combination with direct extracellular flux measurements is supposed to be the most powerful method for obtaining intracellular flux information with only a few modeling assumptions on the living system.

The focus of this contribution will be on tracer experiments in combination with the NMR measurement technique since bioprocess engineers are usually rather unfamiliar with these methods. The reader is referred to [4, 19, 20] for more details on metabolite balancing. Throughout the text main emphasis will be on modeling and data analysis. The mathematical tools introduced will then be used for investigating the general potential of labeling experiments for getting information about the living system. Only those properties of NMR are discussed that are required for understanding the origin of the data sets used for flux analysis. More details on in vivo NMR can be taken from [21, 22, 24, 25] while recent developments in reaction engineering for in vivo NMR are reviewed in [26]. An illustrative application example concerned with the whole central metabolism of *Corynebacterium glutamicum* will conclude the text. The biological implications of the presented results are discussed in [18, 27].

1.3 Data Sources for Stationary Flux Analysis

Reliable physiological data can only be obtained by measuring methods that do not influence the living system. The most familiar source of information is given by all quantitatively relevant extracellular fluxes like substrate consumption, product formation, biomass growth or gas efflux. These fluxes can easily be obtained from standard bioreactor instrumentation [28] and analytical procedures using simple mass balancing.

The other source of data is presented by the isotopic label distribution in intermediates and products obtained from tracer experiments. Briefly, a metabolic carbon isotope tracer experiment is carried out by replacing a substrate (e.g. glucose) with a substrate that is ^{13}C or ^{14}C labeled at a certain carbon atom position. From this moment on, the label is distributed over the whole network until, finally, the fractional labeling in all carbon atoms of intracellular metabolites equilibrates. In other words, after running through an *isotopically nonstationary* state the system finally reaches an *isotopically stationary* state but always remains in a *metabolically stationary* state.

It depends on the experimental setup and the measurement technique whether the isotopically nonstationary state can be observed or only the final isotopically stationary state. Figure 8 in [26] illustrates the dynamic progress of an isotopically nonstationary ^{13}C tracer experiment with respect to labeling as observed within an NMR instrument while Table 2 in this contribution is based on stationary data.

Carbon tracer experiments are the most often used isotope labeling experiments for quantitative flux determination although ^{2}H and ^{15}N tracers or tracer combinations have also been applied [29–33]. Among the carbon isotopes ^{13}C has become the most popular because it can easily be detected with an NMR instrument. For this reason we concentrate here on ^{13}C labeling.

Evaluating isotopically nonstationary tracer experiments requires the additional knowledge of intracellular metabolite pool sizes. In some cases these can already be derived from additional NMR measurements, or otherwise they have to be measured from cell extracts (cf. [26]). If both data sources are not available it may be possible to estimate the pool sizes as well as the unknown fluxes from the measured data by parameter-fitting (see Sect. 4.7).

1.4 Some Typical Experimental Setups

The resulting data set for flux quantitation is illustrated in Fig. 1. If it is sufficiently large, all intracellular fluxes can be quantitated based on very few assumptions on the living system as will be shown below. Some basic types of experiments for achieving this goal can now be distinguished where, in many cases, the experiment has been repeated with differently labeled input substrates [34–36]:

1. Only extracellular fluxes are measured [4, 5, 10].
2. Only labeling fractions in an isotopically stable state are measured [12, 14, 37].
3. Extracellular flux measurements are combined with labeling fractions from an isotopically stationary state [16–18, 34, 38].
4. Extracellular flux measurements and the time course of label enrichment in an isotopically nonstationary state is observed [39–42].
5. Extracellular flux measurements and so called isotopomer fractions (see Sect. 1.6) in an isotopically stationary state are available [15, 43– 45].

1.5 Preconditions for Stationary Flux Analysis

As mentioned before, stationary flux determination by tracer experiments combined with extracellular flux measurements relies on only a few modeling assumptions. However these should be explicitly mentioned.

1. The system is in a metabolic stationary state during the time span taken by the experiment. Clearly this can be established inside a modern controlled bioreactor operated in a continuous culture mode (e.g. in turbidostatic or nutristatic culture).
2. For the metabolic pathways of interest, all reaction steps of the underlying biochemical network must be known with respect to the biochemical reactions involved and the fate of all carbon atoms within each step. For the central metabolism this knowledge is well established and can be taken from any biochemistry text book.
3. It is assumed that enzymes make no difference between labeled and unlabeled species of their substrates. Of course this is the basic assumption for all kinds of tracer studies. However, it should be mentioned that some small molecules have been shown to exhibit isotope mass effects under certain conditions [46–48].

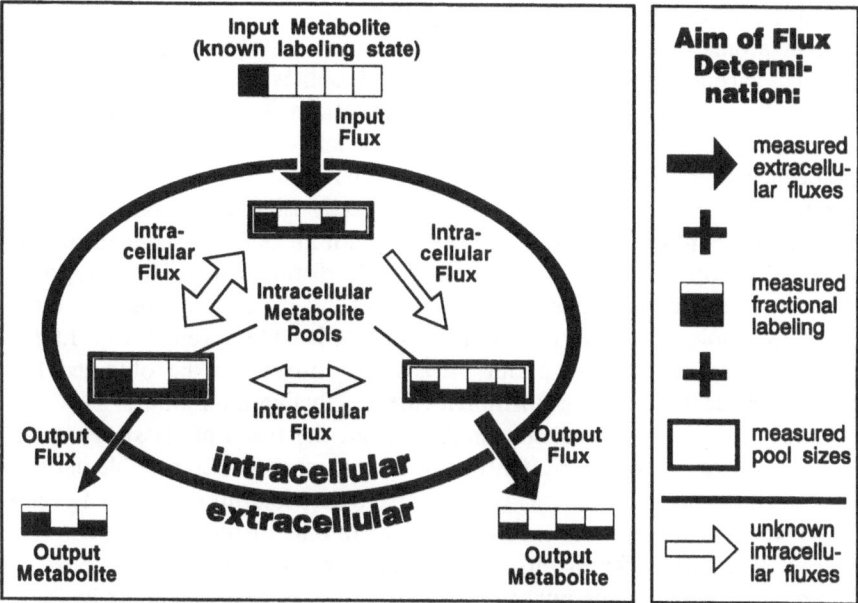

Fig. 1. Available measurement data for stationary flux analysis: extracellular fluxes, fractional carbon labeling, metabolite pool sizes

4. The measurement process does not influence the cell function. This holds true for standard online instruments and modern sampling techniques with rapid cell inactivation if only small samples are taken [26]. On the other hand, although it is generally assumed that strong magnetic fields do not affect microbial metabolism, this is questioned by a recent publication [49]. However using the measurement procedure of [50] the organisms are not actually cultivated inside the NMR instrument so that this assumption is no longer necessary (cf. Sect. 2.4).
5. In the case where analysis of cellular material is performed after the experiment (e.g. from cell extracts or whole inactivated cells) it must be assured that the measured data is representative of the in vivo state of the system (cf. Sect. 2.4).

1.6 Isotopomers

Up to now it is not clear why ^{13}C labeling has outgrown the classical ^{14}C technique over the last years. The reason is that – apart from the intrinsic problems of working with radioactive material – much more information can be obtained much more easily by using the NMR technique. This will be explained now in some more detail (cf. [24, 51]).

The main problem of tracer quantitation is to distinguish between labeled carbon atoms at different positions within one metabolite. To this end, in classical carbon isotope approaches, the metabolites had to be extracted and chemically degraded for separating the single carbon atom positions [34, 45, 52]. In contrast to this time-consuming procedure, an NMR instrument allows one to localize directly all ^{13}C labeled atoms at the same time within a mixture of substances. This is even possible within intact cells using in vivo NMR (cf. [26]).

Even more information can be obtained with an NMR instrument because so-called isotopomers can – at least in part – be distinguished. The isotopomers of a metabolite with n carbon atoms represent the 2^n possible labeling states in which this molecule can be encountered (Fig. 2). It will be seen in Sect. 1.8 that, in fact, more information about intracellular fluxes can be obtained from isotopomer data than from positional carbon labeling data alone.

Clearly, isotopomer measurement is beyond the reach of classical methods based on chemical treatment. Only mass spectroscopy is also capable of distinguishing between isotopomers. Applications to flux determination are described in [43, 45, 53]. However, mass spectrometry can only distinguish between those isotopomers with different numbers of labeled carbon atoms while more isotopomer fractions can be quantitated with NMR (cf. Sect. 4.6 and Fig. 2).

1.7 The Significance of Fractional Carbon Labeling Data

Clearly, extracellular flux data is directly related to intracellular fluxes by stoichiometric balance equations. On the other hand, it is not clear a priori that

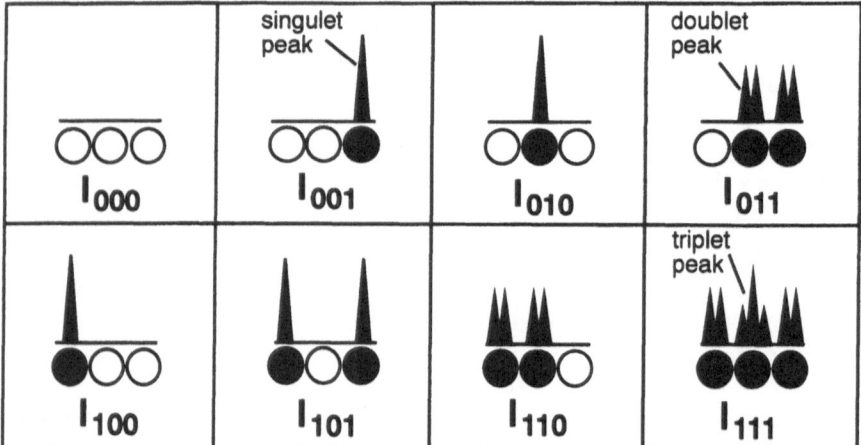

Fig. 2. The $2^3 = 8$ different ^{13}C-isotopomers of an ideal molecule I with three carbon atoms. *White circles* correspond to ^{12}C, *black circles to* ^{13}C. When identical scalar coupling constants between adjacent carbon atoms are assumed (ideal case) the peak patterns in the corresponding high resolution NMR spectra are schematically shown

labeling data contains any information on fluxes. A very simple example shown in Fig. 3a makes clear that there are indeed strong relations between fluxes and fractional labels, even in isotopic equilibrium.

The example shows a network with two alternate parallel pathways for the formation of a product P from a substrate S, both with two carbon atoms. The alternate pathways via the intracellular metabolites A and B are distinguished by the different fate of the carbon atoms from S. If a [2-^{13}C]-labeled substrate is fed into the system (see Fig. 3a), two different isotopomers of the product will emerge, one labeled at the first and the other at the second carbon position.

Assume now that the fluxes in the alternate pathways are given by v_A, v_B and the percentage of labeling for both carbon atoms of P are given by p_1, p_2. From this we get

$$\frac{v_A}{v_B} = \frac{p_1}{p_2}.$$

Consequently the flux ratio can be determined from the label fraction ratio. A similar situation is encountered within *Corynebacterium glutamicum* which produces lysine using two different pathways [54]. The above argument was used in [37] to quantitate the usage of these different pathways (see also [14]).

Clearly this information cannot be calculated from any direct extracellular flux measurement. On the other hand the example reveals the general property of tracer experiments that without (absolute) extracellular flux measurements only flux ratios can be computed. If in the example the product formation

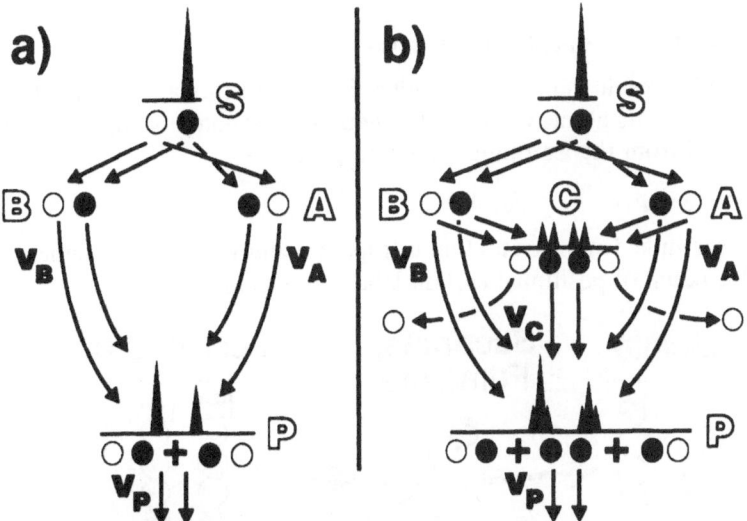

Fig. 3a, b. Example networks demonstrating the superiority of tracer experiments for flux determination compared to methods based on extracellular fluxes alone: **a** two alternate pathways for producing P can be distinguished based on positional carbon labels; **b** three pathways can be distinguished based on isotopomers by multiplet measurement

$v_P = v_A + v_B$ is known in addition we get (notice that $p_1 + p_2 = 1$ because labeling is conserved in the system)

$$v_A = p_1 v_P \quad v_B = p_2 v_P.$$

Of course there would be no measurable effect if the fate of the label were the same in both branches. As a rule of thumb, the flux information that can be obtained from labeling experiments depends heavily on the extent to which position-changing of carbon atoms occurs. Fortunately, it happens quite frequently that labeled carbon atoms are distributed over the complete metabolic network. This finally explains why carbon tracers are the most promising isotopes for intracellular flux determination while e.g. ^{15}N can only be used for special investigations [29].

1.8 The Significance of Fractional Isotopomer Data

The example can be further extended for demonstrating that isotopomer measurements can reveal even more details about flux distributions. In Fig. 3b another product formation step has been added that produces P from A and B via a bimolecular reaction step and an intermediate metabolite. With the indicated [2-^{13}C] substrate labeling, this reaction forms a third isotopomer of P with both positions labeled.

The additional flux is denoted by v_C and the isotopomer percentages in P are given as indicated in Fig. 3b by p_{01}, p_{11}, p_{10} (note that the unlabeled isotopomer does not occur). We then have $p_{01} + p_{10} + p_{11} = 1$ and $v_P = v_A + v_B + v_C$ so that

$$v_A = p_{10} v_P \quad v_B = p_{01} v_P \quad v_C = p_{11} v_P.$$

Of course this result cannot be produced with positional carbon atom labeling measurements alone. On the other hand the carbon labeling state can be reconstructed from the isotopomer knowledge as

$$p_1 = p_{10} + p_{11} \quad p_2 = p_{01} + p_{11}$$

and thus the example demonstrates that isotopomer analysis is always superior or at least the equal of positional carbon label analysis (cf. Sect. 5.4).

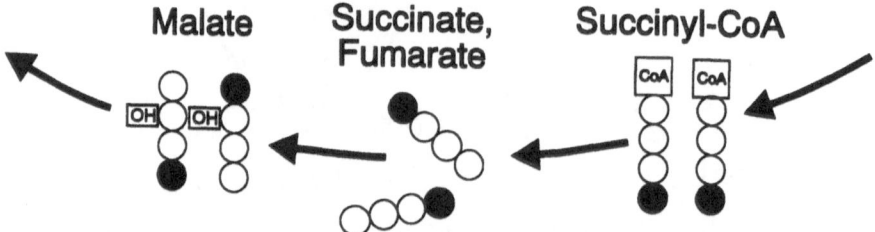

Fig. 4. Label scrambling in the citric acid pathway. The symmetric molecules of succinate and fumarate can freely change their orientation so that labeled carbon atoms finally occur in two positions of malate with equal probability

Clearly, the presence of a bimolecular reaction step is essential for producing this effect. A similar situation is encountered in the citric acid cycle where carbon atoms can change positions within scrambling reactions (Fig. 4). This basically is the phenomenon that is used in [15, 55] for quantifying citric acid fluxes from isotopomer measurements.

1.9 Relations Between Stationary and Kinetic Modeling

A short comparison between stationary flux analysis and the investigation of the dynamic behaviour of metabolic fluxes in response to changing extracellular conditions will conclude this section. Both approaches differ strongly with respect to the experimental procedures and measurement equipment used as well as the results that can be achieved. On the other hand their common goal is to obtain a quantitative characterization of biochemical reaction steps in vivo. In both situations, very recently several powerful new experimental techniques have been developed (cf. [26]).

With respect to mathematical modeling, dynamic investigations always aim at the development and experimental validation of detailed kinetic models of intracellular metabolism (e.g. [56, 57]). On the other hand, for the metabolically stationary situation, much simpler models are required to compute unknown intracellular fluxes from stationary measurement data. There are some inter-relations between the stationary and the dynamic approach showing that both are really complementary to each other.

- Kinetic models usually approximate the dynamics of metabolism around a certain stationary state [57, 58]. Consequently, stationary analysis supplies the cornerstones for dynamic modeling.
- No assumptions on kinetic mechanisms have to be made for stationary modeling. Consequently, the results of stationary flux determination are expected to be more reliable than those from mechanistic modeling.
- As will be shown in Sect. 4.4 a distinguishing feature of stationary tracer experiments is their potential for quantifying both directions of a bidirectional reaction step under certain conditions [59] (Fig. 5). This may be valuable information for distinguishing equilibrating reaction steps from controlling reaction steps in vivo. Moreover, for mechanistic modeling this has the consequence that reversible enzymatic mechanisms have to be taken into consideration [57, 61, 62].
- Of course, results from stationary flux analysis are only valid around the investigated metabolic state [63]. On the other hand a validated mechanistic model may expose a larger prediction horizon and may even be capable of forecasting the effect of genetic modifications. However, in both cases a series of experiments under different conditions has to be performed in order to obtain a complete picture.

2 Measuring Stationary Intracellular Data

2.1 Extracellular Flux Data

Almost all stationary studies make use of extracellular flux measurements to a certain degree. If a bioreactor is used for cell cultivation the extracellular fluxes can be calculated using mass balancing from concentration measurements and the dilution rate. For example, HPLC or FIA instruments, gas efflux measurements and the determination of biomass concentration can be used for this purpose (cf. [26]).

An important idea was the incorporation of cell mass composition for the quantitation of the anabolic fluxes that use precursors from central metabolism [64, 65]. As shown by [66] the biosynthetic pathways of any cell component can be uniquely traced back to 12 precursors in central metabolism. These are glucose-6-phosphate, fructose-6-phosphate, ribose-5-phosphate, erythrose-4-phosphate, glyceraldehyde-3-phosphate, 3-phosphoglycerate, phosphoenolpyruvate, pyruvate, acetyl-coenzyme A, α-ketoglutarate, succinyl-coenzyme A and oxaloacetate. From this knowledge, the knowledge of cell composition and the determined biomass growth rate, a detailed quantitation of the corresponding 12 effluxes from central metabolism is obtained which dramatically improves the available information.

For convenience, many experiments rely on the assumption that biomass composition in microorganisms is constant over a large variety of metabolic states and microorganisms [65, 67]. Of course this assumption requires great care and ought to be experimentally verified from time to time, though this is a very laborious task.

2.2 Some more Details on NMR Spectra

In order to understand the following modeling considerations, some more details on NMR spectra are now briefly sketched at the risk of oversimplification. The reader is referred to [21, 22, 24–26] for an in-depth discussion of in vivo NMR measurement techniques.

NMR spectra are superpositions of resonance spectra from all resonating atoms within the sample. In principle each ^{13}C atom produces a unique resonance peak in the NMR spectrum. Its frequency position depends on the

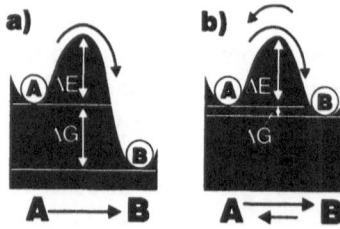

Fig. 5a, b. Uni- and bidirectional reaction steps from the thermodynamic viewpoint. It depends on the in vivo free reaction energy ΔG and the activation energy ΔE whether a reaction proceeds in only one direction or in both directions simultaneously

electro-chemical surroundings of this atom in the metabolite it is part of. The area of the corresponding resonance peak is directly proportional to the concentration. An example spectrum taken from an in vivo experiment is given in Fig. 8 of [26].

In most cases the ^{13}C isotopes are not directly observed but ^{1}H NMR is used instead as an indirect measuring technique. The ^{13}C isotopes then become detectable by a changed resonance pattern of the surrounding protons since their ^{1}H peaks split up into peaks from ^{12}C-bound protons and peaks from ^{13}C-bound protons. This enables the relative amount of labeled carbon atoms to be determined without knowing the absolute amount of the examined substance. Due to the low chemical shift dispersion and broad complex multiplet structures, ^{1}H spectra are much harder to interpret than ^{13}C spectra. For this reason metabolites usually have to be purified before they can be effectively measured with ^{1}H NMR [37]. An example spectrum is given in Fig. 6.

This situation becomes even more difficult when high resolution spectra are taken for detecting isotopomers. The underlying phenomenon is that, due to scalar spin-spin-coupling, a ^{13}C resonance peak corresponding to some carbon atom splits up into a so-called multiplet peak when adjacent atoms are also labeled. The number of peaks in the multiplet depends on the number of labeled neighbours so that doublets, triplets, quartets and so on can be observed. It

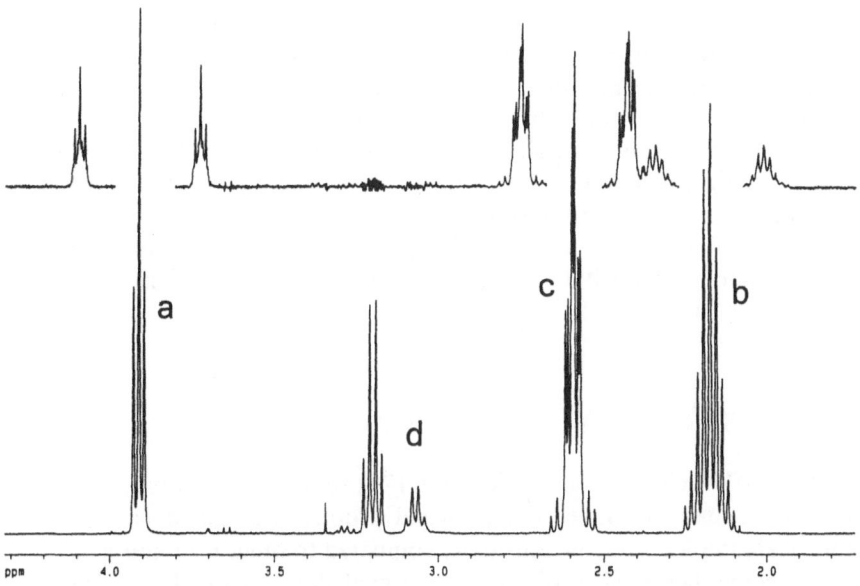

Fig. 6. Proton NMR spectra of ^{13}C labeled glutamate extracted from protein of *C. glutamicum* after incubation with [1-^{13}C] glucose. *Lower trace*: ^{13}C-decoupled spectrum showing a) H-2, b) H-3, c) H-4 protons of glutamate and d) signals from impurities. *Upper trace*: corresponding ^{13}C satellite position signals obtained after subtracting the ^{13}C-decoupled spectrum from the "normal" proton spectrum as explained in [37]

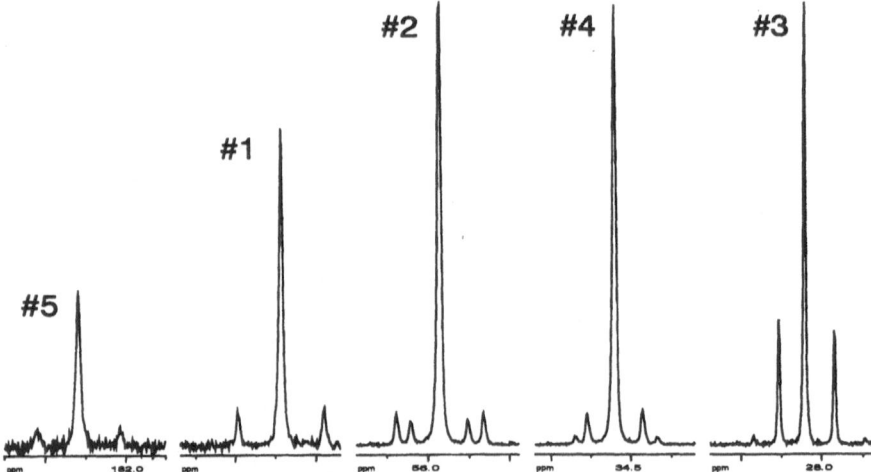

Fig. 7. ^{13}C isotopomer spectra of the ^{13}C labeled glutamate shown in Fig. 6. Multiplets are due to scalar coupling with adjacent ^{13}C atoms (cf. [44])

depends on the specific molecule which spectra will actually result from isotopomers. Figure 2 shows schematically for an ideal C-3 body how in principle the spectral peaks correspond to certain labeling states of the observed molecules while Fig. 7 shows what results from a real metabolite (glutamate). However the exact correspondence between isotopomers and spectra in most cases is known in advance from published NMR data (e.g. [68]).

It now becomes clear, why ^{13}C does not always reliable to completely separate between all possible isotopomers (compare to [44, 45, 69]). For example in Fig. 2 a mixture of the isotopomers I_{100} and I_{001} will produce singlet peaks on the first and third position from which both isotopomer fractions can be quantitated. On the other hand, if I_{101} is also present in the mixture the two peaks cannot be uniquely attributed to the three isotopomers (cf. Sect. 4.6).

2.3 Approaches for NMR of Biological Samples

In vivo NMR techniques are reviewed in [26] so that only the most important facts are summarized here. The principal problem associated with NMR of biological samples is its low sensitivity compared to other techniques like mass spectroscopy. The quality of an NMR signal depends on the measurement duration for producing the spectrum and the amount of labeled material within the sample. For isotopomer quantitation, high resolution spectra are required while a lower resolution may be sufficient for determining only positional carbon enrichment. If the isotopically nonstationary state has to be observed the measurement duration for taking one spectrum is strongly limited, resulting in generally low signal to noise ratios. These problems have been partially overcome in the last few years by the development of more powerful NMR instruments, the increased availability of ^{13}C labeled substrates and the experimental techniques that are described in the following.

The strongest in vivo NMR signals are obtained when the volume of the NMR receiver coil is completely occupied by cellular material as is approximately the case in many studies on perfused organs [15,45]. In this situation ^{13}C NMR spectra with acceptable signal to noise ratios can be produced in less than a minute for metabolites in the millimolar concentration range. This allows time courses [42] and even isotopomers to be observed [15,44].

In contrast to mammalian organs, microorganisms expose much higher oxygen demands and faster growth rates. As a consequence, even in hollow fiber bioreactors well suited for high density cultivation of mammalian cells, they cannot be maintained in such high concentrations and in a well defined reproducible and stable physiological state (cf. [26]). One way to overcome these problems was to take concentrated cell suspensions [12,14,70]. However it is doubtful if such results are representative for the in vivo state.

The best systems presently available for in vivo NMR studies of microorganisms under truly well defined stationary conditions in continuous culture are specially developed continuous flow NMR bioreactors [71,72]. In such reactors cell densities of up to $30–50\,gl^{-1}$ (dry cell mass) can be maintained and measured on line with in vivo NMR. These systems are very well suited for monitoring time courses of ^{13}C label incorporation [26]. On the other hand the spectral quality obtained is still not optimal for isotopomer determination.

When mixtures of labeled substances are measured – as is always the case with whole cells – a great many signals of different compounds in widely differing concentrations can be found in the spectrum. In particular, for low resolution spectra the peaks may significantly overlap [73]. This poses the problem of spectral deconvolution for disentangling the resonances and computing the peak areas. Several numerical methods have been developed to treat this problem [74–77].

2.4 Decoupling Production of Labeled Material from Measuring

An important idea for improving the NMR signal quality was to decouple the biological labeling experiment from the NMR measuring process. Taking samples and extracting intracellular metabolites is generally insufficient because most substances are at far too low a concentration (usually below $1\,mmol\,l^{-1}$) to produce an NMR signal. A successful approach is to take advantage of the cell anabolism, i.e. of the fact that most intermediates are finally stored in cellular components like protein, lipids, RNA or DNA. In this form they are hidden from NMR observation but can be extracted by hydrolization in combination with preparative analytic measures (Fig. 8). Several authors report the

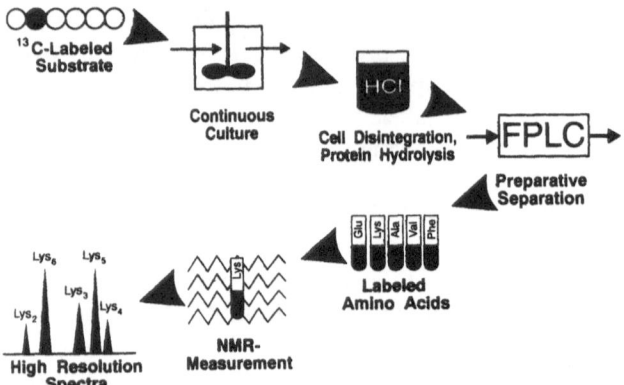

Fig. 8. Decoupling production of labeled material from measuring by protein hydrolyzation after cultivation. Amino acid fractions are obtained by preparative separation methods

extraction of glycerol from lipids [35, 36, 78], ribonucleotides from RNA/DNA [78, 79] and amino acids from protein [50, 78, 80, 81].

The fractions obtained can be kept in the NMR instrument for an arbitrarily long time so that high resolution spectra and even multiplet peaks can be observed [80]. It is a great advantage of the stationary approach that no absolute pool concentrations are required (cf. Sect. 4.7) and mass deficiencies occurring during the separation process do not change the labeling fractions in the sample. On the other hand the isotopically nonstationary states cannot be reconstructed with this technique.

Clearly, the largest amount of quantitative information can be taken from the amino acids because they are synthesized from precursors at so many different positions in the central metabolic network. From their labeling state the fractional enrichment of their precursors can be immediately derived. Most of these precursor pools are inaccessible with other methods because they are too lowly concentrated in vivo. This technique has been used first in a quantitative manner in [18] which, at the same time, is the first application of the tracer technique to a continuously cultivated microorganism.

As pointed out in Sect. 1.5, it must be assured that the measured data is representative of the in vivo state of the system. In particular the isotopically steady state of the system has to be guaranteed [79, 82]. This is a nontrivial problem since, at the time when the intermediary metabolic pools have reached an isotopically stationary state after switching to the labeled substrate, this still does not hold for the cell components. The reason is that the cell mass in the bioreactor is originally unlabeled. However, after several cell residence times, unlabeled cell material is washed out until finally the fractional labeling in the cell protein is representative of the in vivo state. On the other hand only a limited number n of residence times can be awaited for financial reasons, which can be corrected by the *washout correction factor* [18]

$$\text{WCF} \overset{def}{=} \frac{1}{1 - e^{-n}}. \tag{1}$$

3 Modeling of Metabolic ^{13}C Labeling Systems

3.1 Modeling Frameworks for Tracer Experiments

In order to evaluate the data set from a ^{13}C labeling experiment a model has to be formulated that quantitatively describes the relations between fluxes and ^{13}C labels. This model can then be fitted to the measured extracellular fluxes and intracellular fractional labels.

Carbon isotope labeling studies fall within the general category of tracer experiments. General models and mathematical tools for tracer analysis have already been developed in the 1970s; an excellent textbook is [83]. However there are some new aspects in metabolic carbon isotope labeling systems compared to general linear tracer systems.

- In "classical" applications of tracer experiments the model contains only a few pools (usually less than ten) and only one or two of them could be observed [84]. This is compensated for by the availability of isotopically nonstationary observations (i.e. time courses). In the case of metabolic systems the situation is frequently quite different because a larger number of pools can be observed but only in the isotopically stationary state.
- Each metabolic reaction step induces several carbon atom transitions that take place with the same reaction rate. This allows one to impose further constraints on metabolic tracer systems that are not given in the general case.
- The intracellular labeling state can be independently influenced by the forward and backward flux of a bidirectional reaction step (see Sect. 4.4). Although this effect is well known for general tracer systems [83,84] it has seldom been taken into account for metabolic networks to larger extent [34, 36, 42, 85]. Thus, bidirectional steps should be conceptionally integrated into a general modeling framework [59,86].
- Isotopomer dynamics cannot be handled within the framework of linear tracer kinetics because quadratic equations are required to describe the system (see Sect. 3.11). On the other hand they still expose some general structure that can be formalized using a concise matrix notation from [87].

In the context of network modeling and data evaluation it is advisable to set up a general modeling framework for studying arbitrary nets and to facilitate variational studies. A so-called structural system representation using matrix notation for separating network properties from flux and label variables greatly facilitates the development of general mathematical tools for simulation and systems analysis [88, 89]. However, only in a few cases has a general structural modeling approach been taken [18, 90, 92]. The modeling framework presented below is based on the general linear tracer model [82, 83] while general modeling of exchange rates is done as in [18] and basic equations for isotopomer balancing taken from [87].

3.2 Constructing the Biochemical Network

Constructing a model for labeling systems requires knowledge of all biochemical reaction steps in the network under consideration and, moreover, the fate of all carbon atoms within each reaction step. A simple formal notation first introduced in [93] is used here to represent this information. It is derived from the familiar chemical sum notation presented in [4] and can be used later for automatic model generation. As an example the transaldolase step in the pentose phosphate pathway is written as

$$
\begin{aligned}
\text{TA: GAP} \quad &+ \text{ S7P} \qquad > \text{E4P} \quad + \text{F6P} \\
\#\text{ABC} &+ \#\text{abcdefg} > \#\text{defg} + \#\text{abcABC}.
\end{aligned}
$$

This means that the first carbon atom of GAP (denoted by the sign # and the capital letter A) is taken over to the fourth carbon atom of F6P and so on.

Some bimolecular reaction steps require consideration of two molecules of one substance. This is conveniently expressed by including this substance twice and denoting the duplicate carbon atoms with different symbols. For example, the conversion of fructose 1,6-bisphosphate to glyceraldehyde 3-phosphate is simply written as

$$
\begin{aligned}
\text{ALD: F16BP} \quad &> \text{GAP} \quad + \text{GAP} \\
\#\text{ABCDEF} &> \#\text{CBA} + \#\text{DEF}.
\end{aligned}
$$

Scrambling reactions (Fig. 4) are usually assumed to be symmetric, i.e. both scrambling steps have equal probability. Thus the situation shown in Fig. 4 can easily be expressed by

$$
\begin{aligned}
\text{SCR: SUCCCOA} + \text{SUCCCOA} &> \text{MAL} \qquad + \text{MAL} \\
\#\text{ABCD} \quad + \#\text{UVWX} \quad &> \#\text{ABCD} + \#\text{XWVU}.
\end{aligned}
$$

3.3 Noninteger Stoichiometric Coefficients

Sometimes non-integer coefficients have been used to represent further knowledge on metabolic networks [4, 67]. The most important example is biomass composition that can be represented by a biomass formation "reaction". However, such equations make no sense when carbon atoms have to be traced through the network. Fortunately, biomass production can be represented by a set of equations that describes the incorporation of each singular precursor metabolite into biomass. For example, incorporation of glucose 6-phosphate into biomass is written as

G6PBM: G6P   > G6PBiomass
      #ABCDEF > #ABCDEF.

The corresponding "non-integer coefficient" is then supplied separately as a flux measurement value for G6PBM computed from biomass composition and growth rate. This is in any case the more natural way to account for biomass composition.

When no labeling data is available the measured carbon fluxes are usually insufficient to determine the unknown intracellular fluxes from the stoichiometric balance equations. In this situation further assumptions have to be made [65], and other types of balances have to be considered in addition. As an example of an additional assumption, some enzymes like the malic enzyme have been assumed inactive in [65]. On the other hand, flux balancing is extended by considering energy metabolism (i.e. ATP, NADH or NADPH) in [4, 5, 10, 67].

Usually the production of ATP from NADH is assumed to be a reaction step with known stoichiometry. However, it should be noticed that oxidative phosphorylation is not based on a mechanism with fixed stoichiometry and only a few facts are known of its thermodynamic efficiency in vivo [67, 94]. Moreover the possibility of futile cycling makes NADH, NADHP and ATP balancing a delicate problem [35, 95–99]. The same holds for direct energy balancing based on free reaction energies [67] which may depend heavily on the physiological situation, i.e. ΔG^0 values cannot be directly taken over to the in vivo situation (cf. [100] and Fig. 5).

3.4 An Example System

In the following a simple example system taken from [87] is used for introducing the various balances that have to be formulated. It is chosen to demonstrate several general features of labeling systems with as few metabolites as possible. The reaction steps and carbon atom transitions in the system are given in the introduced formal notation by

v1: A   > B   v5: B   > K
    #12   > #12   #12 > #12

v2: B   > E   v6: C   > D   + F
    #12   > #12   #1234 > #234 + #1

v3: B   + E   > C   v7: D   > E   + G
    #12 + #34 > #1234   #123 > #12 + #3

v4: E   > H
    #12   > #12

Three networks are associated with this system [102]. The metabolite network is shown in Fig. 9a, the carbon network in Fig. 9b and a small section of the isotopomer network in Fig. 9c. It becomes clear that, even for small systems, isotopomer networks can become quite complex. For this reason a "full sized" section from central metabolism is completely unsuitable as an example.

However, the example network bears a resemblance to the citric acid cycle in connection with the anaplerotic reaction section. It should be noticed that carbon atoms change their position after one turn in the citric acid cycle while they remain unchanged within an anaplerotic step. These properties are displayed by the example network too, but much less carbon atoms are required.

3.5 Simplification of the Example System

To further reduce the number of isotopomers in the example system it is now transformed to an equivalent simpler system shown in Fig. 10. The idea behind this simplification is to backtrack the product carbon atoms F#1 and G#1 through the system. It obviously, makes no difference if these atoms are already split off from B#1 and E#2 instead of C#1 and D#3. This allows one to remove C#1, D#3 and C#4 from the system. Finally, G and C can be completely removed because nothing splits off at these stages. The remaining

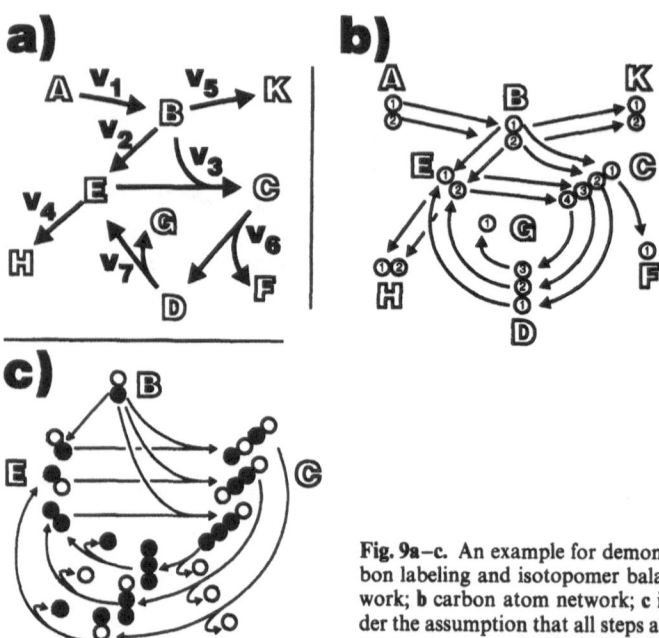

Fig. 9a–c. An example for demonstrating metabolite, carbon labeling and isotopomer balancing: **a** metabolite network; **b** carbon atom network; **c** isotopomer network under the assumption that all steps are unidirectional and the input is labeled as indicated. In this situation only 13 of the 48 possible isotopomers are really produced.

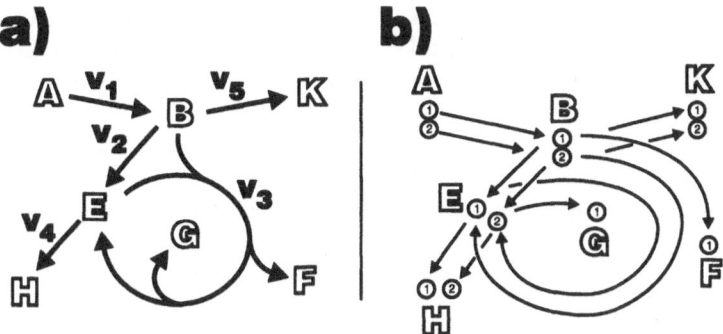

Fig. 10a, b. Reduced network from Fig. 9 as a result of network simplification: **a** metabolite network; **b** carbon atom network

network is now composed from v1, v2, v4, v5 and the changed reaction step

$$v3: \quad B \quad + \quad E \quad > \quad E \quad + \quad F \quad + \quad G$$
$$\#12 \quad + \quad \#34 \quad > \quad \#23 \quad + \quad \#1 \quad + \quad \#4.$$

Although this network looks strange, all quantities connected with the original network shown in Fig. 9 can be reconstructed from the following computations based on the reduced system. This emphasizes that network reduction is an important task whenever isotopomers have to be accounted for. The considerations made above give some impression of the typical simplification operations used in the literature to reduce network complexity.

3.6 Flux State Variables

In the following, the values of the metabolic fluxes $v1, ..., v5$ are denoted by corresponding variables $v_i \geq 0$. More precisely, two variables $v_i^{\rightarrow}, v_i^{\leftarrow}$ have to be introduced representing each forward and backward flux direction (Fig. 5). Clearly, if a reaction step is irreversible, one of these fluxes is zero. In the example it will be assumed that the uptake and product formation steps v1, v4 and v5 are irreversible (a familiar modeling assumption) as well as the intracellular step v3. On the other hand v2 is assumed to take place in both directions.

In order to comprise the forward and backward directions of all molar fluxes in the reaction network the vectors

$$\mathbf{v}^{\rightarrow} = (v_1^{\rightarrow}, v_2^{\rightarrow}, v_3^{\rightarrow}, v_4^{\rightarrow}, v_5^{\rightarrow}) \quad \text{and} \quad \mathbf{v}^{\leftarrow} = (v_1^{\leftarrow}, v_2^{\leftarrow}, v_3^{\leftarrow}, v_4^{\leftarrow}, v_5^{\leftarrow})$$

are introduced. The additional irreversibility assumptions are given by

$$v_1^{\leftarrow} = v_3^{\leftarrow} = v_4^{\leftarrow} = v_5^{\leftarrow} = 0.$$

The physical unit of fluxes is strictly taken to be $[\mathrm{mol}\,h^{-1}]$. The reason is that a chemical reaction step does not only take over substrates to products but also substrate carbon atoms (respectively isotopomers) to product carbon atoms (respectively isotopomers). Clearly, if the unit $[\mathrm{mol}\,h^{-1}]$ is chosen, the same flux values $v_i^{\rightarrow}$, $v_i^{\leftarrow}$ can serve for representing metabolite fluxes as well as carbon or isotopomer fluxes.

Finally, from the vectors $\mathbf{v}^{\leftarrow}$, $\mathbf{v}^{\rightarrow}$ the composite overall flux vector

$$\mathbf{v} = \begin{pmatrix} \mathbf{v}^{\rightarrow} \\ \mathbf{v}^{\leftarrow} \end{pmatrix}$$

(of twice the dimension) and the corresponding net flux vector

$$\mathbf{v}^{net} = \mathbf{v}^{\rightarrow} - \mathbf{v}^{\leftarrow}$$

are formed (notice that with standard measurement equipment only some of these net fluxes can be observed). Clearly, all components of $\mathbf{v}$ must be non-negative which is expressed by the component-wise inequality

$$\mathbf{v} \geq \mathbf{0}.$$

3.7 Pool State Variables

The labeling state is always represented by the percentage of labeled material at all carbon atom positions. Only the intermediary metabolites and the metabolites fed into the system have to be accounted for balancing. For denoting the label fractions within a metabolite like B, small indexed letters b_1, b_2 are used. In general all intermediary carbon atom pools under consideration have to be enumerated. This gives rise to the fractional label variables $x_i \in [0, 1]$ and the overall labeling vector $\mathbf{x}$. In our case, only B and E are intermediates such that

$$\mathbf{x} = (b_1, b_2, e_1, e_2)^T.$$

A special type of carbon atoms are those that are fed into the system as a metabolic substrate because their labeling state is known a priori. These atoms are enumerated likewise and form the constant input labeling vector

$$\mathbf{x}^{inp} = (a_1, a_2)^T.$$

When isotopomers are considered the situation is much more complex. We distinguish between the isotopomer fractions of one metabolite by using a binary number that specifies, which carbon atoms are labeled [103]. For instance the $2^2 = 4$ isotopomer fractions of B are denoted by b_{00}, b_{01}, b_{10}, b_{11}. Since carbon atoms and isotopomers cannot be confused in the following, the symbol $\mathbf{x}$ is used again to denote the isotopomer fraction state vector

$$\mathbf{x} = (b_{00}, b_{01}, b_{10}, b_{11}, e_{00}, e_{01}, e_{10}, e_{11})^T \tag{2}$$

and the relations

$$b_{00} + b_{01} + b_{10} + b_{11} = 1 \quad b_{10} + b_{11} = b_1 \quad e_{10} + e_{11} = e_1$$
$$e_{00} + e_{01} + e_{10} + e_{11} = 1 \quad b_{01} + b_{11} = b_2 \quad e_{01} + e_{11} = e_2 \qquad (3)$$

always hold between isotopomer and carbon label fractions. As in the carbon atom case, the vector $\mathbf{x}^{inp}$ of input isotopomers has to be defined as

$$\mathbf{x}^{inp} = (a_{00}, a_{01}, a_{10}, a_{11})^T.$$

Finally, the modeling of isotopically nonstationary experiments requires the knowledge of absolute molar pool sizes. For a specific metabolite they are denoted by capital italic letters like B for the pool size of B. In the general model the corresponding pool variables are related to the vector

$$\mathbf{X} = (B, E).$$

3.8 Metabolite Balance Equations

Several types of material balances can now be formulated for each intracellular pool using the introduced state variables. The first is the well known stoichiometric balance equation [19] that holds for the fluxes participating in one metabolite pool. Because the absolute pool sizes do not change in a metabolic stationary state, the sums of incoming and outgoing fluxes must be equal. In the example the stoichiometric balance equations corresponding to the intermediary metabolite pools are

$$B: v_1^{\rightarrow} + v_2^{\leftarrow} = v_2^{\rightarrow} + v_3^{\rightarrow} + v_5^{\rightarrow}$$
$$E: v_2^{\rightarrow} + v_3^{\rightarrow} = v_2^{\leftarrow} + v_3^{\rightarrow} + v_4^{\rightarrow}. \qquad (4)$$

By introducing the stoichiometric matrix

$$\mathbf{N} = \begin{pmatrix} 1 & -1 & -1 & . & -1 \\ . & 1 & . & -1 & . \end{pmatrix}$$

this can be more conveniently expressed as

$$0 = \mathbf{N} \cdot \mathbf{v}^{net}. \qquad (5)$$

3.9 Stationary Carbon Label Balance Equations

When carbon atom labeling is considered, a carbon label balance can be written for each intermediary carbon pool. In this situation the above reaction system should be interpreted as a system of carbon atom transitions

v1: $A\#1 > B\#1$	v3: $B\#1 > F\#1$	v4: $E\#1 > H\#1$
v1: $A\#2 > B\#2$	v3: $B\#2 > E\#1$	v4: $E\#2 > H\#2$
v2: $B\#1 > E\#1$	v3: $E\#1 > E\#2$	v5: $B\#1 > K\#1$
v2: $B\#2 > E\#2$	v3: $E\#2 > G\#1$	v5: $B\#2 > K\#2.$

The carbon label balance for B#1 is now constructed as follows. The amount of labeled material that is carried over to B#1 by the incoming flux v1 from A#1 and the backward flux of v2 from E#1 is given by $v_1^\to a_1 + v_2^\leftarrow e_1$. On the other hand the outgoing fluxes v2, v3 and v5 take the amount of $(v_2^\to + v_3^\to + v_5^\to)b_1$ out of B#1. If the complete system is assumed to be in an isotopically stationary state the following set of carbon label balance equations results:

$$b_1: \quad v_1^\to a_1 + v_2^\leftarrow e_1 = (v_2^\to + v_3^\to + v_5^\to)b_1$$

$$b_2: \quad v_1^\to a_2 + v_2^\leftarrow e_2 = (v_2^\to + v_3^\to + v_5^\to)b_2$$

$$e_1: \quad v_3^\to b_2 + v_2^{\to'} b_1 = (v_2^\leftarrow + v_3^\to + v_4^\to)e_1$$

$$e_2: \quad v_3^\to e_1 + v_2^\to b_2 = (v_2^\leftarrow + v_3^\to + v_4^\to)e_2. \tag{6}$$

Again this can be more conveniently expressed using matrix notation as

$$\begin{pmatrix} 0 \\ 0 \\ 0 \\ 0 \end{pmatrix} = \begin{pmatrix} v_1^\to & . \\ . & v_1^\to \\ . & . \\ . & . \end{pmatrix} \begin{pmatrix} a_1 \\ a_2 \end{pmatrix}$$

$$+ \begin{pmatrix} -v_2^\to -v_3^\to -v_5^\to & & & v_2^\leftarrow & & . \\ & -v_2^\to -v_3^\to -v_5^\to & & . & & v_2^\leftarrow \\ v_2^\to & v_3^\to & -v_2^\leftarrow -v_3^\to -v_4^\to & . \\ . & v_2^\to & v_3^\to & -v_2^\leftarrow -v_3^\to -v_4^\to \end{pmatrix} \begin{pmatrix} b_1 \\ b_2 \\ e_1 \\ e_2 \end{pmatrix}.$$

Here the input labels a_1, a_2 are separated from the intermediary labels. Clearly, this equation has the general structure

$$0 = \mathbf{P}^{inp}(v)x^{inp} + \mathbf{P}(v)x = \left(\sum_i v_i \mathbf{P}_i^{inp}\right)x^{inp} + \left(\sum_i v_i \mathbf{P}_i\right)x \tag{7}$$

where $\mathbf{P}_i$, $\mathbf{P}_i^{inp}$ are called the atom transition matrices [59] (compare to the closely related atom mapping matrices in [92]). For example we have

$$P_2^\to = \begin{pmatrix} -1 & . & . & . \\ . & -1 & . & . \\ 1 & . & . & . \\ . & 1 & . & . \end{pmatrix}, \quad P_2^\leftarrow = \begin{pmatrix} . & . & 1 & . \\ . & . & . & 1 \\ . & . & -1 & . \\ . & . & . & -1 \end{pmatrix}, \quad P_1^{inp} = \begin{pmatrix} 1 & . \\ . & 1 \\ . & . \\ . & . \end{pmatrix}.$$

3.10 Nonstationary Carbon Label Balance Equations

If the system is not in isotopic equilibrium, a differential equation system has to be formulated. In this situation the absolute pool sizes X_j play a role because they determine the capacity of a pool for labeled material. In the example the absolute amount of labeled material in the first position of B is given by Bb_1 so that $d/dt(Bb_1) = Bd/dtb_1$ when the system is in a metabolic stationary state. This term has to be added to the label balance equation at stage $B\#1$ in Eq. (6).

In order to get a general matrix representation similar to Eq. (7), a matrix I is constructed that enlarges the vector X to the dimension of x by appropriately repeating its entries. In the example,

$$
\mathbf{IX} \overset{def}{=}
\begin{pmatrix}
1 & \cdot \\
1 & \cdot \\
\cdot & 1 \\
\cdot & 1
\end{pmatrix}
\begin{pmatrix} B \\ C \end{pmatrix}
=
\begin{pmatrix} B \\ B \\ C \\ C \end{pmatrix}.
$$

Putting all parts together we get the nonstationary balance equation system

$$
\mathrm{diag}\,(\mathbf{IX})\frac{d}{dt}\mathbf{x} = \mathbf{P}^{inp}(\mathbf{v})\mathbf{x}^{inp} + \mathbf{P}(\mathbf{v})\mathbf{x} \tag{8}
$$

where $\mathrm{diag}\,(\mathbf{IX})$ is the diagonal matrix constructed from the vector $\mathbf{IX}$.

3.11 Isotopomer Balance Equations

When isotopomers are considered, the situation is more complicated because the number of educts involved in a reaction step determines the (algebraic) order of the reaction equation. In the example the reaction step v3 has two substrates so that at the isotopomer level all reactions

$$
\begin{array}{llllll}
v3: & B\#00 & + \ E\#00 & > \ E\#00 & + \ F\#0 & + \ G\#0 \\
v3: & B\#00 & + \ E\#01 & > \ E\#00 & + \ F\#0 & + \ G\#1 \\
v3: & B\#00 & + \ E\#10 & > \ E\#01 & + \ F\#0 & + \ G\#0 \\
v3: & B\#00 & + \ E\#11 & > \ E\#01 & + \ F\#0 & + \ G\#1 \\
v3: & B\#01 & + \ E\#00 & > \ E\#10 & + \ F\#0 & + \ G\#0 \\
v3: & B\#01 & + \ E\#01 & > \ E\#10 & + \ F\#0 & + \ G\#1 \\
v3: & B\#01 & + \ E\#10 & > \ E\#11 & + \ F\#0 & + \ G\#0 \\
v3: & B\#01 & + \ E\#11 & > \ E\#11 & + \ F\#0 & + \ G\#1 \\
v3: & B\#10 & + \ E\#00 & > \ E\#00 & + \ F\#1 & + \ G\#0 \\
v3: & B\#10 & + \ E\#01 & > \ E\#00 & + \ F\#1 & + \ G\#1
\end{array}
$$

$$v3: \ B\#10 \ + \ E\#10 \ > \ E\#01 \ + \ F\#1 \ + \ G\#0$$
$$v3: \ B\#10 \ + \ E\#11 \ > \ E\#01 \ + \ F\#1 \ + \ G\#1$$
$$v3: \ B\#11 \ + \ E\#00 \ > \ E\#10 \ + \ F\#1 \ + \ G\#0$$
$$v3: \ B\#11 \ + \ E\#01 \ > \ E\#10 \ + \ F\#1 \ + \ G\#1$$
$$v3: \ B\#11 \ + \ E\#10 \ > \ E\#11 \ + \ F\#1 \ + \ G\#0$$
$$v3: \ B\#11 \ + \ E\#11 \ > \ E\#11 \ + \ F\#1 \ + \ G\#1$$

may happen. Clearly, the probability of two isotopomers with fractional amounts b_i and b_j to meet in step v3 is given by $b_i b_j$ [44]. Consequently, the corresponding total isotopomer flux is $v_3^{\rightarrow} b_i b_j$. Thus isotopomer balancing for bimolecular steps leads to bilinear terms with respect to the x_i, while for the monomolecular steps v1, v2, v4 and v5 and the effluxes from each pool all terms in the corresponding balance equations are exactly analogous to those in Eq. (7) and Eq. (8) respectively. The complete equation system then is

$$b_{00}: \ v_1^{\rightarrow} a_{00} + v_2^{\leftarrow} e_{00} = (v_2^{\rightarrow} + v_3^{\rightarrow} + v_5^{\rightarrow}) b_{00}$$
$$b_{01}: \ v_1^{\rightarrow} a_{01} + v_2^{\leftarrow} e_{01} = (v_3^{\rightarrow} + v_2^{\rightarrow} + v_5^{\rightarrow}) b_{01}$$
$$b_{10}: \ v_1^{\rightarrow} a_{10} + v_2^{\leftarrow} e_{10} = (v_3^{\rightarrow} + v_2^{\rightarrow} + v_5^{\rightarrow}) b_{10}$$
$$b_{11}: \ v_1^{\rightarrow} a_{11} + v_2^{\leftarrow} e_{11} = (v_3^{\rightarrow} + v_2^{\rightarrow} + v_5^{\rightarrow}) b_{11}$$
$$e_{00}: \ v_2^{\rightarrow} b_{00} + v_3^{\rightarrow} (b_{00} + b_{10})(e_{00} + e_{01}) = (v_2^{\leftarrow} + v_3^{\rightarrow} + v_4^{\rightarrow}) e_{00}$$
$$e_{01}: \ v_2^{\rightarrow} b_{01} + v_3^{\rightarrow} (b_{00} + b_{10})(e_{10} + e_{11}) = (v_2^{\leftarrow} + v_3^{\rightarrow} + v_4^{\rightarrow}) e_{01}$$
$$e_{10}: \ v_2^{\rightarrow} b_{10} + v_3^{\rightarrow} (b_{01} + b_{11})(e_{00} + e_{01}) = (v_2^{\leftarrow} + v_3^{\rightarrow} + v_4^{\rightarrow}) e_{10}$$
$$e_{11}: \ v_2^{\rightarrow} b_{11} + v_3^{\rightarrow} (b_{01} + b_{11})(e_{10} + e_{11}) = (v_2^{\leftarrow} + v_3^{\rightarrow} + v_4^{\rightarrow}) e_{11}.$$

Again a comprehensive matrix notation is desirable. Clearly, the linear terms in x can still be expressed with the matrix notation introduced in Eq. (7). On the other hand the bilinear terms in x are represented by introducing one symmetric matrix for each balance equation. For example the e_{11} step can be written as follows:

$$e_{11}: \ 0 = \frac{1}{2} v_3^{\rightarrow}
\begin{pmatrix} b_{00} \\ b_{01} \\ b_{10} \\ b_{11} \\ e_{00} \\ e_{01} \\ e_{10} \\ e_{11} \end{pmatrix}^{\!\!\top}
\begin{pmatrix}
\cdot & \cdot & \cdot & \cdot & \cdot & \cdot & \cdot & \cdot \\
\cdot & \cdot & \cdot & \cdot & \cdot & \cdot & 1 & 1 \\
\cdot & \cdot & \cdot & \cdot & \cdot & \cdot & \cdot & \cdot \\
\cdot & \cdot & \cdot & \cdot & \cdot & \cdot & 1 & 1 \\
\cdot & \cdot & \cdot & \cdot & \cdot & \cdot & \cdot & \cdot \\
\cdot & \cdot & \cdot & \cdot & \cdot & \cdot & \cdot & \cdot \\
\cdot & 1 & \cdot & 1 & \cdot & \cdot & \cdot & \cdot \\
\cdot & 1 & \cdot & 1 & \cdot & \cdot & \cdot & \cdot
\end{pmatrix}
\begin{pmatrix} b_{00} \\ b_{01} \\ b_{10} \\ b_{11} \\ e_{00} \\ e_{01} \\ e_{10} \\ e_{11} \end{pmatrix}$$

$$+ \quad v_2^{\rightarrow} b_{11} - (v_2^{\leftarrow} + v_3^{\rightarrow} + v_4^{\rightarrow}) e_{11}.$$

The symmetric matrix in this equation will be denoted by $Q_{3,11}^{\rightarrow}$ indicating the flux number and the target pool. All such matrices $Q_{i,j}$ corresponding to the same flux variable v_i can then be composed to a 3-dimensional matrix structure (i.e. a tensor) Q_i (which may be thought of as a vector of square matrices) and a 3-dimensional-matrix-times-vector product can then be defined as

$$x^T Q_i x = x^T \begin{pmatrix} Q_{i,1} \\ Q_{i,2} \\ \vdots \\ Q_{i,n} \end{pmatrix} x \stackrel{def}{=} \begin{pmatrix} x^T Q_{i,1} x \\ x^T Q_{i,2} x \\ \vdots \\ x^T Q_{i,n} x \end{pmatrix}.$$

Using this notation the isotopomer balance equations can finally be written quite similar to Eq. (7) as [87]:

$$0 = P^{inp}(v)x^{inp} + P(v)x + x^T Q(v)x$$
$$= \left(\sum_i v_i P_i^{inp}\right)x^{inp} + \left(\sum_i v_i P_i\right)x + x^T\left(\sum_i v_i Q_i\right)x. \tag{9}$$

Input 3-dimensional-matrices analogous to P^{inp} are not required because it can be assumed, without loss of generality, that any substrate enters the system via a monomolecular uptake step. It should be clear now how, in principle, reactions with more than two substrates can be represented. However, since such reaction steps can, in practice, be replaced by successive bimolecular reactions, it is unnecessary to introduce matrices with dimension higher than three. Finally, the nonstationary state equations are constructed completely analogous to Eq. (8).

Interestingly, isotopomers have always been considered in connection with the citric acid cycle [13, 43, 44, 103] where essentially only one bimolecular step occurs at the entry point of acetyl-coenzyme A. Since the labeling state of the input isotopomer is known, the equations for the citrate cycle can be written essentially without any truly bilinear term (i.e. $Q_i = 0$ for all i). This reduces the equation system to a linear model with respect to x which can be mathematically treated with the same methods as the ordinary carbon labeling system in Eq. 7 [13].

4 Simulation and Data Analysis

4.1 Network Synthesis

As shown above, all model equations required for describing the isotope and isotopomer labeling system can be built up from certain vectors and matrices. However, the dimensions can become quite large. When the whole central metabolism including glycolysis, pentose phosphate pathway, citric acid cycle, glyoxylate shunt and anaplerotic reaction section is included, the metabolic network has about 25 metabolite fluxes between 20 metabolite pools, 120 carbon fluxes between 80 carbon atom pools and 3200 isotopomer fluxes between 600 isotopomer pools [42, 104]. This makes clear that computer aided tools for network synthesis and consistency checking are absolutely necessary. In particular, model variation studies would be quite time-consuming when using manual input.

Most authors used highly specific programs for simulation or data analysis [36, 44, 51] or general systems based on explicit equation input [41, 55]. Only a few general tools for carbon or isotopomer network synthesis have been designed based on explicit matrix input to generate the system equations [92, 105]. However, for large systems, network synthesis from either matrix or balance equation input is still not satisfying, while for isotopomer systems this effort is almost prohibitive (cf. [13]).

A more convenient way for network synthesis is to write a compiler program for translating a minimal formal input like that presented in Sect. 3.2 into the corresponding matrix structures. For metabolite flux networks several such programs are known [65, 106, 107], while for carbon atom and isotopomer networks the program NMRFlux described in [108] is currently the only implementation. The corresponding algorithms for isotopomer network synthesis are described in [104]. Clearly, the input for other matrix- or equation-based simulation or computer algebra systems can easily be generated once a matrix representation is available [108].

An important feature of a network synthesis program is its ability to check formal consistency conditions. In [102] some criteria have been given that lead to consistent networks. In most cases it is sufficient to check the following criteria for detecting typing errors in the textual input – any carbon atom of an educt must appear exactly once on the product side and vice versa, each molecule must have the same number of carbon atoms within each equation, and input and output metabolites must always occur in a monomolecular reaction.

4.2 Simulation

Simulation of ^{13}C labeling experiments means predicting the outcome of an experiment when all fluxes v are known. For this purpose, values for the components v_i must be given in such a way that the constraints imposed by the stoichiometric equations are respected. Usually the user of a simulation program wishes to fix certain values while others are varied in each simulation run. This process can be supported by appropriate software tools [108].

Assume now that a suitable vector $v \geq 0$ is given. Simulation of ^{13}C labeling experiments then can proceed in different ways.

1. For isotopically nonstationary experiments the ordinary differential equations (see Eq. 8) associated with carbon or isotopomer fluxes has to be solved. For non-stiff systems a higher order Runge-Kutta scheme is well suited [87, 109]. On the other hand, when highly reversing reactions occur the

differential equation system will tend to become stiff. Specialized solvers [13, 110], the preliminary introduction of rapid equilibria by pool lumping [86] or the preliminary reduction of the network size [111] may solve these problems.

2. It is well known that linear tracer systems are globally stable except from some pathological situations, because the corresponding system matrices $\mathbf{P(v)}$ are diagonally dominant [83]. From this the negativity of all eigenvalues can be concluded. For the isotopomer equation system it can in certain cases be shown [87] that its linearization is diagonally dominant in any point $0 \leq \mathbf{x} \leq 1$. Such systems are called dissipative and can also be proven to be globally stable [112]. As a consequence the stationary state can always be computed using a differential equation solver as an iterative procedure.

3. Clearly, when only the stationary solution is of interest, differential equation solving is inefficient because the transient states are of no interest. In particular the treatment of isotopomer systems can be time-consuming because of their high dimensionality. Modifying the well known Euler scheme by introducing relaxation leads to the iterative schemes for linear and nonlinear equation solving [113, 114]. In particular, the Gauss-Seidel algorithm is used in [92] for solving the linear carbon labeling balances and in [87] to solve the isotopomer balances. In any situation the sparsity of the involved matrices can be exploited to speed up the computation.

4. In the case of carbon labeling systems, Eq. (7) can be solved explicitly for the vector $\mathbf{x}$ because $\mathbf{P(v)}$ is invertible (which follows from its diagonal dominance) [83]:

$$\mathbf{x} = \mathbf{x(v)} = -\left(\sum_i v_i \mathbf{P}_i \right)^{-1} \left(\sum_i v_i \mathbf{P}_i^{\text{inp}} \right) \mathbf{x}^{\text{inp}}. \tag{10}$$

It is well known that, up to a dimension of about 100, the iterative solution of linear equation systems cannot compete with direct methods even when sparse matrix representations are used [115]. This turned out to be true for labeling systems too [116]. When a high numerical stability is required, a QR factorization method accompanied by a preconditioner [116] or an explicit monitoring of the condition number [103] is better suited.

4.3 Computing Explicit Solutions

In our example the stationary carbon labeling equations as well as the isotopomer equations can be solved explicitly using computer algebraic methods. For simplicity we henceforth assume that the input substrate is labeled at the second position, i.e.

$$a_1 = 0 \quad a_2 = 1$$

or when expressed with isotopomers

$$a_{00} = 0 \quad a_{01} = 1 \quad a_{10} = 0 \quad a_{11} = 0.$$

To start with the calculation the stoichiometric equations (Eq. 4) are used to eliminate the flux variables $v_3^{\rightarrow}$ and $v_4^{\rightarrow}$:

$$v_3^{\rightarrow} = v_1^{\rightarrow} - v_2^{\rightarrow} + v_2^{\leftarrow} - v_5^{\rightarrow}$$

$$v_4^{\rightarrow} = v_2^{\rightarrow} - v_2^{\leftarrow}.$$

The resulting stationary carbon atom labeling fractions can then be obtained from Eq. (10) as

$$b_1 = v_2^{\leftarrow} \alpha_1 \alpha_2 / \alpha_0$$

$$b_2 = \alpha_3 / \alpha_0$$

$$e_1 = \alpha_1 \alpha_2 \alpha_4 / \alpha_0$$

$$e_2 = \alpha_5 / \alpha_0$$

with the auxiliary terms α_i shown below.

For the isotopomer fractions the computation is much more difficult, and in general impossible, because the corresponding equation set is essentially nonlinear with respect to $\mathbf{x}$. However, in this special case the computer algebraic methods discussed in Sect. 5.3 help to compute the explicit result with the aid of a computer algebra system:

$$b_{00} = 1 - b_{10} - b_{01} - b_{11} \qquad e_{00} = 1 - e_{10} - e_{01} - e_{11}$$

$$b_{10} = v_2^{\rightarrow} v_2^{\leftarrow} \alpha_1 \alpha_2 \alpha_6 / \alpha_0^2 \qquad e_{10} = v_2^{\rightarrow} \alpha_1 \alpha_2 \alpha_4 \alpha_6 / \alpha_0^2$$

$$b_{01} = \alpha_2 \alpha_8 / \alpha_0^2 \qquad e_{01} = v_2^{\rightarrow} \alpha_7 / \alpha_0^2$$

$$b_{11} = v_2^{\leftarrow} \alpha_1^2 \alpha_2^2 \alpha_4 / \alpha_0^2 \qquad e_{11} = \alpha_1^2 \alpha_2^2 \alpha_4^2 / \alpha_0^2$$

with

$$\alpha_0 = v_1^{\rightarrow 3} + v_1^{\rightarrow 2}(3v_2^{\leftarrow} - 2v_5^{\rightarrow}) + v_1^{\rightarrow}(3v_2^{\leftarrow 2} - 4v_2^{\leftarrow} v_5^{\rightarrow} + v_5^{\rightarrow 2})$$
$$\qquad + v_2^{\leftarrow}(v_2^{\leftarrow 2} - 2v_2^{\leftarrow} v_5^{\rightarrow} - v_2^{\rightarrow 2} + v_5^{\rightarrow 2})$$

$$\alpha_1 = v_1^{\rightarrow} - v_2^{\rightarrow} + v_2^{\leftarrow} - v_5^{\rightarrow} \qquad \alpha_2 = v_1^{\rightarrow} + v_2^{\leftarrow} - v_5^{\leftarrow}$$

$$\alpha_3 = \alpha_0 - v_2^{\rightarrow} v_2^{\leftarrow} \alpha_1 \qquad \qquad \alpha_4 = v_1^{\rightarrow} + v_2^{\leftarrow}$$

$$\alpha_5 = \alpha_0 - v_2^{\rightarrow} \alpha_1 \alpha_4 \qquad \qquad \alpha_6 = v_1^{\rightarrow 2} + v_1^{\rightarrow}(2v_2^{\leftarrow} - v_5^{\rightarrow})$$
$$\qquad \qquad \qquad \qquad \qquad + v_2^{\leftarrow}(v_2^{\leftarrow} - v_2^{\rightarrow} - v_5^{\rightarrow})$$

$$\alpha_7 = \alpha_0 \alpha_6 + v_2^{\rightarrow} v_2^{\leftarrow}(2\alpha_6 v_2^{\rightarrow} - \alpha_0 - v_1^{\rightarrow} v_2^{\rightarrow 2})$$

$$\alpha_8 = \alpha_7 - v_2^{\leftarrow} \alpha_1^3 \alpha_4.$$

4.4 Explicit Flux Determination

Assume now that the flux values $v_1^{\rightarrow}$ and $v_5^{\rightarrow}$ can be directly measured and, additionally, the labels b_1, b_2 are available. Then the remaining unknown intracellular fluxes $v_2^{\rightarrow}$, $v_2^{\leftarrow}$ must be determined for reconstructing the whole system state. From the balance equations (Eq. 6) the explicit solutions

$$v_2^{\leftarrow} = v_1^{\rightarrow} \frac{b_1^2}{(b_2 - b_1)(b_2 + b_1 - 1)}$$

$$v_2^{\rightarrow} = (v_1^{\rightarrow} + v_2^{\leftarrow} - v_5^{\rightarrow})\frac{1 - b_2}{b_1} \tag{11}$$

can be computed. The nonlinear mapping

$$\left(\frac{v_2^{\leftarrow}}{v_1^{\rightarrow}}, \frac{v_2^{\rightarrow}}{v_1^{\rightarrow} + v_2^{\leftarrow} - v_5^{\rightarrow}} \right) \overset{1-1}{\longleftrightarrow} \left(\frac{b_1^2}{(b_2 - b_1)(b_2 + b_1 - 1)}, \frac{1 - b_2}{b_1} \right)$$

can thus be used to visualize the correspondence between unknown fluxes and measured labels by a superposition of two contour plots (Fig. 11). A similar technique has been also used in [92] as a graphical tool for flux estimation and sensitivity analysis.

An important observation can be taken from the example. Both directions of the reversible reaction step v2 have been identified from labeling data. This proves once more that ^{13}C NMR labeling experiments are considerably more powerful that experiments that are based solely on metabolite balances. This observation motivates a more detailed consideration of bidirectional reaction steps in the next section.

It turns out that, in the example situation, isotopomer measurements are not necessary for flux determination. However, it is an interesting question, if this is still true, when the measurement $v_5^{\rightarrow}$ is no longer available. In this situation the surplus values b_{11} and e_{01}, e_{10}, e_{11} may contain more information on the unknown fluxes. This question will be answered in Sect. 5.4.

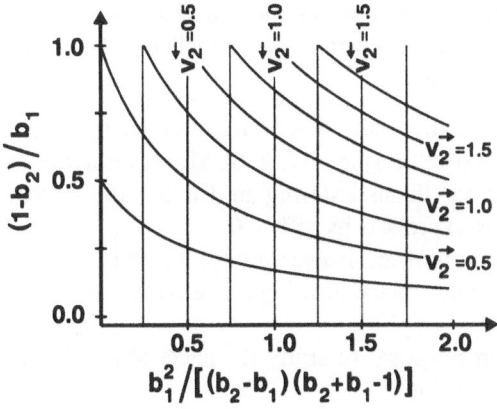

Fig. 11. Superposition of two contour plots illustrating the computation of unknown intracellular fluxes from fractional labeling data. When two extracellular fluxes are assumed to be measured as $v_1^{\rightarrow} = 1.0$ and $v_5^{\rightarrow} = 0.5$, and measurements of b_1, b_2 are available, the unknown fluxes $v_2^{\rightarrow}$ and $v_2^{\leftarrow}$ can be read off from the diagram

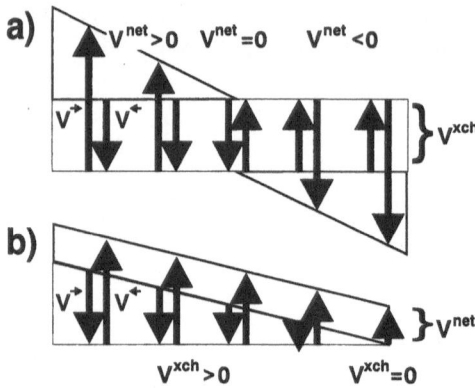

Fig. 12a, b. Definition of exchange fluxes for bidirectional reaction steps showing forward and backward flux: **a** for fixed exchange flux and varying net flux; **b** for fixed net flux and varying exchange flux

4.5 Exchanging Reactions

When doing simulation studies on the influence of exchanging reactions it quickly becomes clear that forward and backward fluxes are well suited for formulating the balance equations (Eq. 7, 9) but rather inconvenient for expressing assumptions on exchange rates. To this end a more suitable coordinate system has to be found in which a forward/backward flux pair $v_i^\rightarrow$, $v_i^\leftarrow$ is described in terms of the net flux rate v_i^{net} and an appropriate exchange flux v_i^{xch}. In [86] the quantity $v_i^{xch} = v_i^\rightarrow + v_i^\leftarrow$ is suggested for investigation of the system's behaviour when $v_i^{xch} \to \infty$ (rapid equilibrium situation). However, this quantity is not convenient for describing irreversibility (i.e. $v_i^\rightarrow = 0$ or $v_i^\leftarrow = 0$) because it depends on v_i^{net} in this situation. Another definition from [18] that serves better for expressing irreversibility assumptions (but is rather inconvenient for analytical purposes) is given by (see also Fig. 12)

$$v_i^{xch} = \min(v_i^\rightarrow, v_i^\leftarrow). \tag{12}$$

It can be easily verified that the pair ($v_i^\rightarrow$, $v_i^\leftarrow$) can be computed from (v_i^{net}, v_i^{xch}) and vice versa. Moreover it should be observed that v_i^{xch} is non-negative but, on the other hand, it does not prescribe a certain net flux direction. Using v^{xch}, physiological assumptions that are frequently made for biochemical reaction steps can be expressed as follows (cf. Fig. 5).

1. Irreversibility assumptions are usually made when large free energy differences are known from the in vitro situation (i.e. $\Delta G^0 \gg 0$). Moreover irreversibility must always be assumed for fluxes entering or leaving the system. Clearly, irreversibility of step i is expressed by $v_i^{xch} = 0$.
2. Rapid equilibrium is the counterpart of irreversibility. In this case the forward and backward reaction takes place at a high rate compared to the net flux rate: $v_i^{xch} \gg v_i^{net}$.
3. Finally it is useful for simulation purposes to study the effect of a v_i^{xch}-variation, by setting v_i^{xch} to arbitrary values.

A simulation run should be parametrized in the $(\mathbf{v}^{net}, \mathbf{v}^{xch})$ coordinate system where $\mathbf{v}^{xch}$ denotes the vector of all exchange fluxes. The stoichiometric equations and the assumptions made on the exchange fluxes impose linear constraints on these coordinates. The formal representation of these constraints can be easily handled by extending the stoichiometric equation (Eq. 5) to a more general linear constraint equation:

$$\mathbf{N}^{cnstr} \begin{pmatrix} \mathbf{v}^{net} \\ \mathbf{v}^{xch} \end{pmatrix} = \mathbf{c}^{cnstr}$$

4.6 Measurement Equations

Incorporating measured values requires the introduction of measurement equations. Again a matrix notation is used where the measurement matrices $\mathbf{M}$ express which components of the state vectors have been measured. The vectors ε denote $N(0, \Sigma)$-distributed noise terms with symmetric and positive definite covariance matrices Σ. Usually the Σ-matrices have diagonal shape, i.e. the measurements are assumed to be independent. With this notation we now get

the flux measurement equation $\qquad\qquad \mathbf{w} = \mathbf{M}_w \cdot \mathbf{v}^{net} + \varepsilon_w$

the label or isotopomer measurement equation $\;\; \mathbf{y} = \mathbf{M}_y \cdot \mathbf{x} + \varepsilon_y$

the pool size measurement equation $\qquad\quad \mathbf{Y} = \mathbf{M}_Y \cdot \mathbf{X} + \varepsilon_Y. \quad (13)$

Clearly, the pool size measurement equation is only required in the isotopically nonstationary case. In this situation the label measurement equation must additionally be extended by a discrete time parameter.

When NMR multiplet analysis is used for isotopomer measurement, the isotopomer measurement matrix expresses how the measured values correspond to the fractional amounts of isotopomers. In the example illustrated by Fig. 2 the observation is composed from singlet peaks s_1, s_2, s_3, doublet peaks d_1, d_2, d_3 and a triplet peak t_2. Moreover the sum of all percentages must be 1. The correspondence between these measured quantities and the isotopomer fractions is given by (cf. [44, 69])

$$\mathbf{M}_y x = \begin{pmatrix} 1 & 1 & 1 & 1 & 1 & 1 & 1 & 1 \\ \cdot & \cdot & \cdot & \cdot & 1 & 1 & \cdot & \cdot \\ \cdot & \cdot & 1 & \cdot & \cdot & \cdot & \cdot & \cdot \\ \cdot & 1 & \cdot & \cdot & \cdot & 1 & \cdot & \cdot \\ \cdot & \cdot & \cdot & \cdot & \cdot & \cdot & 1 & 1 \\ \cdot & \cdot & \cdot & 1 & \cdot & \cdot & 1 & \cdot \\ \cdot & \cdot & \cdot & 1 & \cdot & \cdot & \cdot & 1 \\ \cdot & \cdot & \cdot & \cdot & \cdot & \cdot & \cdot & 1 \end{pmatrix} \begin{pmatrix} i_{000} \\ i_{001} \\ i_{010} \\ i_{011} \\ i_{100} \\ i_{101} \\ i_{110} \\ i_{111} \end{pmatrix} = \begin{pmatrix} 1 \\ s_1 \\ s_2 \\ s_3 \\ d_1 \\ d_2 \\ d_3 \\ t_2 \end{pmatrix} = y.$$

Interestingly this M_x is not a full rank matrix but there remains only one degree of freedom to determine all isotopomer fractions from the measurements. Its null space is generated by the vector $(1, -1, 0, 0, -1, 1, 0, 0)^T$, i. e. the isotopomers $I\#000, I\#100, I\#001$ and $I\#101$ cannot be separated (compare to Sect. 2.2).

On the other hand, if mass spectroscopy is used isotopomers can only be distinguished by their total mass so that measured quantities m_0, m_1, m_2, m_3 are obtained. We then have

$$
\begin{pmatrix}
1 & 1 & 1 & 1 & 1 & 1 & 1 & 1 \\
1 & . & . & . & . & . & . & . \\
. & 1 & 1 & . & 1 & . & . & . \\
. & . & . & 1 & . & 1 & 1 & . \\
. & . & . & . & . & . & . & 1
\end{pmatrix}
\begin{pmatrix}
i_{000} \\ i_{001} \\ i_{010} \\ i_{011} \\ i_{100} \\ i_{101} \\ i_{110} \\ i_{111}
\end{pmatrix}
=
\begin{pmatrix}
1 \\ m_0 \\ m_1 \\ m_2 \\ m_3
\end{pmatrix}.
$$

This is obviously much less information, which explains why NMR is, in principle, superior to mass spectroscopy. However, both methods can be combined with the time-consuming chemical degradation technique which enables a complete isotopomer analysis to be performed [45, 53].

4.7 Flux Estimation

In the general situation, flux estimation cannot be achieved explicitly as in Sect. 4.4 or graphically using the graphical superposition technique as demonstrated in Fig. 11, because the number of unknown parameters is too large for reasonably complex networks (cf. Sect. 4.1). In this situation a nonlinear regression approach using the familiar least squares estimator is appropriate [82, 117]. Knowing that x is always a function of v from Eq. (10), this flux estimate is obtained as the solution of the nonlinear programming problem

$$
\text{minimize } \kappa(v) = \|w - M_v v^{net}\|_{\Sigma_w}^2 + \|y - M_x x(v)\|_{\Sigma_y}^2
$$

$$
\text{constrained by } N^{cnstr} \begin{pmatrix} v^{net} \\ v^{xch} \end{pmatrix} = c^{cnstr} \quad \text{and} \quad v \geq 0 \tag{14}
$$

where $\|\xi\|_\Sigma^2 = \xi^T \Sigma^{-1} \xi$ denotes the squared weighted norm corresponding to a covariance matrix Σ. In the case where only flux measurements are available, this is exactly the linear estimate for flux estimation from extracellular flux data proposed in [4] that can be directly computed using generalised inverse matrices [118].

In the isotopically nonstationary state the situation becomes more complex because, usually, not all pool sizes can be measured. Consequently, they also have to be estimated from the measured data, i.e. $\kappa(\mathbf{v})$ in Eq. (14) now becomes $\kappa(\mathbf{v}, \mathbf{X})$ and the sum of squares has to be extended by the term $\|\mathbf{Y} - \mathbf{M}_x\mathbf{X}\|^2_{\Sigma_Y}$.

This shows that nonstationary experiments require more information compared to the stationary case. This problem is usually treated by assuming fixed sizes for all small pools from literature data [40–42] while taking measurements for the large pools [41]. This strategy seems to be justified because the model outcome is often very insensitive with respect to small pool sizes [41, 42].

4.8 Solution of the Flux Estimation Problem

In the majority of applications the flux estimation problem is solved explicitly as has been done in Sect. 4.4. However this approach cannot in general make use of the complete measurement information. The same holds for the graphical approach (Fig. 11) that is strongly limited to dimensions one and two. On the other hand the general numerical solution of the flux estimation problem (Eq. 14) poses several problems.

1. The linear constraints have to be resolved. This can be done with appropriate numerically stable matrix factorization techniques like singular value decomposition [65, 119].
2. Exchange fluxes as defined by Eq. (12) are only piecewise differentiable functions. This problem can be treated with a derivative free algorithm like the well known Nelder Mead simplex algorithm at the cost of large computation times as has been done in [18].
3. The inequality constraints $\mathbf{v} \geq \mathbf{0}$ have to be strictly obeyed. An ad hoc solution to overcome this problem is simply to replace $\mathbf{v}_i$ by some λ_i^2 which is always positive and then to minimize over λ_i. However, this will significantly decrease numerical stability.
4. While in older papers the minimum of $\kappa(\mathbf{v})$ has been found manually by trial and error [34, 39] iterative optimization algorithms are now used [18, 41, 44]. A modern successive quadratic programming algorithm that can simultaneously handle the constraints is developed in [116].
5. In many cases the computational complexity of the nonstationary flux estimation problem has been overcome by more or less dramatical simplifications of the network [40, 42, 86, 120].

4.9 Statistical Analysis

Statistical analysis is required to judge the quality of the measured data and the obtained estimates. Several well established statistical methods can be used for this purpose.

- The ability of the model to describe the measured data set can be tested [35, 41, 103, 121].

- Redundancies in the measured data set can be used to detect measurement errors [65, 122, 123].
- Sensitivity analysis of the model output with respect to the input parameters is used to study their influence [36, 42, 55, 89, 111].
- The sensitivity of the estimated parameters with respect to the measured quantities shows how the estimates are influenced by single measurements [87].
- The approximate covariance matrix for the estimated parameters can be computed and from this approximate parameter confidence regions can be constructed [35, 44, 87].

All sensitivities as well as the covariance matrix can be computed when the derivative of $x(v)$ from Eq. (7) with respect to v is known. It can be computed by implicit differentiation as follows [89]:

$$0 = \left(\sum_i v_i P_i\right)\frac{\partial x}{\partial v_j} + (P_j x + P_j^{inp} x^{inp}).$$

Thus $\partial x/\partial v$ can be computed using the same matrix factorization that has already been used for computing $x(v)$ in Eq. (10). This further emphasizes the use of direct methods instead of an iterative solution of Eq. (7) (cf. Sect. 4.2). A similar implicit differentiation formula can be proven for the isotopomer system (Eq. 9) [87].

Finally, it should be mentioned that sensitivity analysis is only an approximative approach to statistical analysis because the originally nonlinear model is replaced by its linearization. It is well known that this can lead to serious extrapolation errors [124, 125]. Moreover it can be shown [87, 116] that this effect will most likely occur when large exchange fluxes are estimated. In the case where a graphical method is applicable (cf. Sect. 4.8), nonlinear confidence regions can be immediately derived from the graphical representation [55, 87, 92]. A more general approach to estimate nonlinear parameter confidence regions that uses nonlinear statistical methods is developed in [87].

5 Global Analysis of Stationary Labeling Systems

5.1 Problems of Global System Analysis

This section concentrates on the principal amount of information that can be obtained from metabolic carbon labeling experiments. In this context the results of a parameter-fitting procedure are always unsatisfactory for various reasons.

1. Parameter-fitting produces local results, i.e. a global optimum can never be guaranteed. In [93] an example from the pentose phosphate pathway has been given that admits for two alternative flux solutions both with good

(local) statistical quality measures. An even more complicated example is given in [92] where either one or two solutions can occur in a certain system state.

2. The result is an a posteriori result, i.e. it cannot be decided in advance (i.e. a priori) if the measurements will contain sufficient information for flux determination.
3. The measurement of fractional labels or even isotopomers is a time-consuming procedure. If redundancies in this input data can be predicted a priori this will save a lot of time because some measurements do not have to be performed.
4. Any additional a priori characterization of the outcome of an experiment can be used for complexity reduction. This is of great importance when isotopomer systems are considered.

Obviously, these questions are of great importance for the design and evaluation of experiments. In the case of flux analysis from extracellular flux measurements alone, we are confronted with a linear system for which all questions posed above can be explicitly and efficiently answered [4, 11, 122]. On the other hand, label balancing introduces algebraic equations to the system so that more advanced methods have to be used for system analysis. Interestingly many questions can be answered in this situation too.

5.2 Identifiability and Redundancy

The questions posed above are better known as identifiability and redundancy problems in control engineering [126]. In this context the central problems are:

identifiability a posteriori, which means that all fluxes v can be uniquely determined from a given data set (w, y);

identifiability a priori, which means that all fluxes v can be uniquely determined whatever the outcome (w, y) of the experiment will be; and

redundancy of measurements, which means that there exist relations $f(w, y) = 0$ that hold independently of the non-measurable fluxes in v.

In any case the measurements must be assumed to be taken without error (i.e. $\varepsilon_w = \varepsilon_y = 0$). This means no restriction because sensitivity can be studied later by using the methods presented in Sect. 4.9.

It may be observed that Eqs. (8) when combined with Eq. (13) represents a general parametrized linear state space model with measured variables y. For such models many results on identifiability have been proven [127]. On the other hand the stationary case of Eq. (7) has never been considered explicitly because the number of measurements was too low to obtain significant results. This makes stationary flux identification essentially a new problem.

5.3 Algebraic Methods

Explicit and sometimes quite long-winded algebraic calculations for deriving explicit flux solutions of metabolic carbon labeling systems can be found in numerous publications [15, 16, 55, 128, 129]. In each case the solution strategy is highly application specific and based on various simplifying assumptions on the network structure. The usual assumption is that most reaction steps are either irreversible or in rapid equilibrium [16, 17]. Moreover, whole metabolic pathways like the pentose phosphate pathway are found to be lumped to one reaction step [16, 130]. If any new equation is inserted in the system all computations have to be reworked and, most likely, a completely new solution strategy has to be found. This will almost surely happen when exchanging reactions are added because the complexity of networks is in a close relation to the number of cyclic pathways they contain [131, 132].

A general approach to algebraic identifiability analysis presented in [123] and [89] is based on network simplification algorithms [133, 134] and computer algebraic methods [135, 136] (see [103] for a more empirical approach). The general idea is to derive automatically explicit equations relating fluxes to label measurements. In particular one can try to express the unknown fluxes in terms of measured quantities as has been done in Sect. 4.4. Similar computer algebraic algorithms for multivariate polynomial equations have been successfully applied to identifiability problems in control engineering [137], solution of stationary biochemical reaction systems [138] or stationary optimization of processes [139]. The details would exceed the scope of this text so that only the results from the example are given here.

5.4 Analysis of the Example System

For the example system it has been already proven by Eq. (11) that all fluxes are identifiable a priori if $v_1^{\rightarrow}, v_5^{\rightarrow}$ and b_1, b_2 are known. Clearly, it follows the identifiability a posteriori. On the other hand the following redundancy relations can be proven to hold independent of the actual fluxes $v_1^{\rightarrow}, v_2^{\rightarrow}, v_2^{\leftarrow}, v_5^{\rightarrow}$:

$$0 = b_1 e_1 - b_2^2 - b_1 + b_2 \qquad 0 = b_2 - b_2^2 - b_1 + b_{11}$$

$$0 = b_1 e_2 - b_2 e_1 - b_1 + e_1 \qquad 0 = b_1 e_{11} - e_1 b_{11}.$$

From this we immediately obtain

$$e_1 = (b_2^2 + b_1 - b_2)/b_1 \qquad b_{11} = b_2^2 + b_1 - b_2$$

$$e_2 = (b_2 e_1 + b_1 - e_1)/b_1 \qquad e_{11} = b_{11} e_1/b_1$$

while b_{00}, e_{00} follow from Eq. (3). Consequently, b_1 and b_2 contain all the information on fluxes that can be obtained by label measurements. All the other label as well as isotopomer fractions are redundant! This example shows that isotopomer measurements need not increase the available information on intracellular fluxes as was the case in the example from Sect. 1.8.

6 Application to *Corynebacterium glutamicum*

6.1 Example Organism and Measured Data

Corynebacterium glutamicum has always been of great interest in amino acid production, which is closely coupled to the central metabolic pathways, i.e. glycolysis, pentose phosphate pathway and citric acid cycle. Thus stationary flux analysis is a promising diagnostic method in the context of metabolic engineering for amino acid production. The *C. glutamicum* strain MH20-22B studied in [18] is known as a lysine producer. Since, in this text, the focus is on the principles of flux analysis, more details on the biology of this organism can be taken from [27]. The following results are taken from [18].

C. glutamicum MH20-22B was cultivated under lysine producing conditions with a dilution rate of 0.1 h^{-1} in continuous culture. In this situation the total substrate uptake rate was 1.49 mmol g^{-1} h^{-1} of dry cell mass from which 18.3% lysine was obtained. Table 1 presents all measured extracellular fluxes that are normed to a 100% substrate influx for convenience. For calculating the biomass effluxes a biomass composition similar to that of [66] was assumed.

Intracellular fractional labels were measured using the decoupling technique described in Sect. 2.4. To obtain a (nearly) equilibrated labeling state in the protein fraction three cell residence times were taken for incubation with [1-^{13}C] glucose corresponding to a washout correction factor of 1.05 (cf. Eq. 1). The separated amino acids and the corresponding label enrichment can be taken from Table 2. All measured NMR spectra are of high quality, and an example is shown in Fig. 7. From these spectra a measurement error below the values given in Table 2 can be asserted.

Table 1. Extracellular fluxes measured in continuous culture of *C. glutamicum*. All values are normed to a 100% glucose uptake rate of 1.49 mmol g^{-1} h^{-1} (dry cell mass) Metabolite abbreviations can be taken from Fig. 13 and are assumed to be self explaining (see [18] for details)

Flux	Measured Value [%]	Flux	Measured Value [%]
Substrate uptake:			
GLC	100.0		
Biomass effluxes:		Biomass effluxes:	
G6P	1.3	RI5P	1.0
F6P	0.5	RI5P[b]	4.9
GAP	0.9	AKG	7.0
PYR	18.0	AKG[b]	1.2
PYR[a]	23.0	OAA	11.6
E4P[a]	1.8	Product formation:	
		LYS[a]	18.3
		CO2	275.1

[a] Flux coupled to CO$_2$ formation
[b] Flux coupled to CO$_2$ refixation

Table 2. Some fractional labels measured from protein hydrolysate of *C. glutamicum* compared to values predicted by the balance equations with the estimated fluxes [18]. The measurement precision depends on the quality of the measured spectra. Labeled CO_2 was measured by mass spectroscopy

Carbon Atom	Measured Value [%]	Estimated Value [%]	Measurement Precision [%]	Carbon Atom	Measured Value [%]	Estimated Value [%]	Measurement Precision [%]
E4P#1	2.0	2.5	1.0	OAA#2	7.7	9.8	2.0
E4P#2	3.6	2.0	1.0	OAA#3	20.9	22.6	2.0
E4P#3	2.0	1.9	1.0	OAA#4	16.8	17.3	2.7
E4P#4	16.7	15.3	2.0	LYS#2	6.8	7.1	0.2
GAP#1	2.9	2.7	0.2	LYS#3	21.9	24.0	0.3
GAP#2	2.6	2.6	0.1	LYS#4	18.9	17.3	1.0
GAP#3	26.7	26.3	0.2	LYS#5	22.2	24.9	1.0
PYR#2	3.0	2.7	1.0	LYS#6	5.6	5.3	0.3
PYR#3	26.4	26.3	0.5	CO2#1	23.0	21.6	0.4
AKG#2	24.1	22.6	0.3				
AKG#3	11.1	9.8	0.5				
AKG#4	28.1	26.3	0.6				

6.2 Biochemical Network

The detailed biochemical reaction equations in the formal notation introduced in Sect. 3.2 are not reproduced here and only the underlying metabolite network used for flux estimation is presented in Fig. 13. Reaction steps that have been assumed to be bidirectional are labeled therein by an additional box with rounded edges. Irreversibility assumptions were made on the basis of thermodynamic considerations. The scrambling steps from Fig. 4 have been introduced in the citric acid cycle. Both lysine production pathways [54] have been incorporated.

Comment has to be made on the anaplerotic reaction section (cf. [65]). There are three possible anaplerotic reaction steps catalyzed by PEP carboxylase, PEP carboxykinase and the malic enzyme while even a fourth enzyme (pyruvate carboxylase) may be present [140]. Only the PEP carboxykinase step is supposed to be reversible. The identifiability of the corresponding fluxes from label measurements has been discussed in [89]. It turned out that for the anaplerotic section only a combined net flux from the lumped phosphoenolpyruvate-pyruvate pool to the oxaloacetate-malate pool can be estimated accompanied by an exchange flux. Interestingly, more details of anaplerotic fluxes would be identifiable if the malate labeling state were available.

6.3 Achieved Results

The intracellular flux estimates (Eq. 14) computed from the measured data are shown in Fig. 13. Moreover the simulated labeling state corresponding to the estimates is presented in Table 2. The table shows that all label measurements are well reproduced by the simulation run. The measured extracellular fluxes are

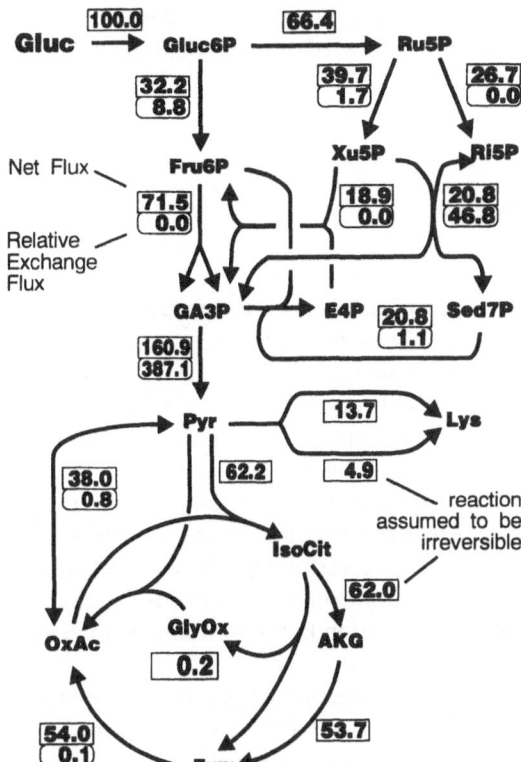

Fig. 13. Biochemical network used for flux analysis in *Corynebacterium glutamicum* MH20-22B under lysine producing conditions (data from [18]). Estimated stationary net fluxes are given in *rectangular boxes* while the associated exchange fluxes for bidirectional reaction steps given in *rounded boxes* are taken relative to the corresponding net fluxes, i.e. the values v^{xch}/v^{net} are represented. Effluxes to biomass (see Table 1) have been left out for simplicity

even better reproduced but are not given here for reasons of space. From computed statistical quality measures given in [116] it becomes clear that all net fluxes are well determined. On the other hand the estimated exchange fluxes are only accurate within an order of magnitude (compare to Sect. 4.9). However, it is possible to decide if a reaction step is highly reversible or rather unidirectional, which is the principal goal of the analysis.

Some remarkable facts that are closely linked to modeling and the general stationary flux determination problem shall be pointed out. Further biological implications are discussed in [27].

1. Bidirectional reaction steps frequently occur and can be quantitated in vivo, which seems to be impossible in vitro [141]. Consequently, the consideration of such steps in the modeling process is absolutely necessary. This in turn requires a large amount of measurement data for estimating all unknown parameters.
2. All fluxes are estimated without the incorporation of energy balances. Thus results on the cellular energy metabolism can be derived from the quantitative results achieved. For example it turns out that there is an excess NADHP formation for which a consuming reaction is not yet known [18]. A balanced NADPH formation was previously assumed in [65].

3. The exchange rate of the anaplerotic reaction section indicates a futile cycle which once more stresses the importance of bidirectional reactions (cf. [96, 99]).

7 Conclusion and Future Prospects

Stationary flux determination is proven to be an invaluable diagnostic tool in the context of metabolic engineering. When sufficient measurement data is available it can be carried out with almost no critical assumptions on the living system like, e.g. energy balancing. For this purpose fractional labeling data from carbon tracer experiments is an important source of information in addition to direct extracellular flux measurements. Using this source enables not only net fluxes to be estimated but also exchange fluxes of bidirectional reaction steps. These in turn allow one to distinguish between equilibrating and irreversible reaction steps in vivo. This was illustrated successfully for the complete central metabolism of *C. glutamicum* in [18]. If isotopomer measurements are available this may even increase the amount of information but it need not be so in every case.

From general modeling considerations it becomes clear that as many information sources as possible should be used for flux determination. In this context isotopomer measurements are a promising source of information that is currently not extensively used. Modeling and data analysis with isotopomer systems will require additional efforts for solving the associated high dimensional numerical problems. The same holds for the statistical analysis of the estimated parameters which poses a difficult nonlinear statistical problem [116]. Finally, from the viewpoint of global system analysis, general methods for network reduction, identifiability and redundancy analysis have to be developed that allow one to judge the amount of information that can be achieved with a certain experiment. Finally, appropriate software tools for stationary flux analysis are required when this technique is to be established in interdisciplinary research teams [108, 142].

From the experimental viewpoint the measurement procedures have to be further accelerated to establish stationary flux analysis by ^{13}C tracer experiments as a routine procedure. Only the evaluation of a series of experiments under varied physiological conditions can bring a true insight into metabolic regulation [4–6]. Similarly, the comparison of different strains which are distinguished by well known genetic modifications will demonstrate the role of a certain enzymatic step within a complex network. Finally, stationary flux analysis – being free from assumptions on the biological system to a large degree – may help one to find out how enzymes really work in vivo, i.e. phenomena like channeling, enzyme complexes or scrambling [143, 144] may be investigated in more detail when sufficient measurement information is available.

8 References

1. Bailey JE (1991) Science 252: 1668
2. Stephanopoulos G, Sinskey AJ (1993) TibTech 11: 392
3. Kacser H, Acerenza L (1993) Eur J Biochem 216: 361
4. Vallino JJ, Stephanopoulos G (1993) Biotech Bioeng 41: 633
5. Jorgensen H (1995) Biotech Bioeng 46: 117
6. Sonntag K, Schwinde J, de Graaf AA, Marx A, Eikmanns BJ, Wiechert W, Sahm H (1995) Appl Microbiol Biotech (In Press)
7. Kacser H (1988) in: Bazin MJ, Prosser JI (ed) Physiological Models in Microbiology, vol 2, CRC Press, p 1
8. Kell DB, Westerhoff HV (1986) FEMS Microbiol Rev 39: 305
9. Albe KR, Wright BE (1992) J Biol Chem 267: 3106
10. Goel A, Ferrance J, Jeong J, Ataai MM (1993) Biotech Bioeng 42: 686
11. van Heijden RTJM, Heijnen JJ, Hellinga C, Romein B, Luyben KCAM (1994) Biotech Bioeng 43: 3
12. Walker TE, Han CH, Kollman VH, London RE, Matwiyoff NA (1982) J Biol Chem 257: 1189
13. Chance EM, Seeholzer SH, Kobayashi K, Williamson JR (1983) J Biol Chem 258: 13785
14. Walsh K, Koshland DE (1984) J Biol Chem 259: 9646
15. Malloy CR, Sherry AD, Jeffrey FMH (1988) J Biol Chem 263: 6964
16. Sharfstein ST, Tucker SN, Mancuso A, Blanch HW, Clark DS (1994) Biotech Bioeng 43: 1059
17. Zupke C, Stephanopoulos G (1995) Biotech Bioeng 45: 292
18. Marx A, de Graaf AA, Wiechert W, Eggeling L, Sahm H (1995) Biotech Bioeng (In press)
19. Hofmeyr JHS (1986) Comp Appl Biosc 2: 5
20. Varma A, Palsson BO (1994) Bio/Technol 12: 994
21. Gadian DG (1982) Nuclear Magnetic Resonance and its Application to Living Systems. Oxford University Press, Oxford
22. Matwiyoff NA (1982) in: [23] p 573
23. Schmidt H-L, Förstel H, Heinzinger K (ed) (1982) Stable Isotopes, Elsevier (Analytical Chemistry Symposia Series, vol 11)
24. London RE (1988) Progr NMR Spectr 20: 337
25. Lundberg P, Harmsen E, Ho C, Vogel HJ (1990) Anal Biochem 191: 193
26. Weuster D, de Graaf AA (1995) Adv Biochem Eng Biotech This volume
27. Eggeling L, de Graaf AA (1995) Adv Biochem Eng Biotech This volume
28. Schügerl K (ed) (1991) Measuring, Modelling and Control Verlag Chemie, Weinheim (Biotechnology, vol 4)
29. Kanamori K, Weiss RL, Roberts D (1988) J Biol Chem 263: 2871
30. Roberts MF, Choi B-S, Robertson DE, Lesage S (1990) J Biol Chem 265: 18207
31. London RE (1992) in: Berliner LJ, Reuben J (ed) Biological Magnetic Resonance, vol 11, Plenum Press, p 277
32. Malaisse WJ, Biesemans M, Willem R (1994) Mol Cell Biochem 130: 129
33. Ross BD, Kingsley PB, Ben-Yoseph O (1994) Biochem J 302: 31
34. Stein RB, Blum JJ (1979) J Biol Chem 254: 10385
35. Rabkin M, Blum JJ (1985) Biochem J 225: 761
36. Crawford JM, Blum JJ (1983) Biochem J 212: 595
37. Sonntag K, Eggeling L, de Graaf AA, Sahm H (1993) Eur J Biochem 213: 1325
38. Jans AWH, Winkel C, Buitenhuis L, Lugtenburg J (1989) Biochem J 257: 425
39. Kelly P, Kelleher JK, Wright BE (1979) Biochem J 184: 581
40. Fitzpatrick SM, Hetherington HP, Behar KL, Shulman RG (1990) J Cerebr Blood flow Metab 10: 170
41. Weiss RG, Gloth ST, Kalil-Filho R, Chacko VP, Stern MD, Gerstenblith G (1992) Circ Res 70: 392
42. Chatham JC, Forder JR, Glickson JD, Chance EM (1995) J Biol Chem 270: 7999
43. Katz J, Wals P, Lee W-NP (1993) J Biol Chem 268: 25509
44. Künnecke B, Cerdan S, Seelig J (1993) NMR in Biomed 6: 264
45. Di Donato L, Des Rosiers C, Montgomery JA, David F, Garneau M, Brunengraber H (1993) J Biol Chem 268: 4170

46. Wolfsberg M (1992) in: [23] p 3
47. O'Leary MH (1982) in: [23] p 67
48. Winkler FJ, Kexel H, Kranz C, Schmidt H-L (1982) in: [23] p 83
49. Okuno K (1994) J Ferment Bioeng 77: 453
50. Marx A (1994) Charakterisierung des Zentralstoffwechsels bei *Corynebacterium glutamicum* mittels Metabolitbilanzierung und computergestützter Analyse von ^{13}C-NMR-Markierungsdaten Diploma Thesis, Universität Bonn
51. Jeffrey FMH, Rajagopal A, Malloy CR, Sherry AD (1991) TIBS 16: 5
52. Cohen SM, Rognstad R, Shulman RG, Katz J (1981) J Biol Chem 256: 3428
53. Inbar L, Lapidot A (1987) Eur J Biochem 162: 621
54. Schrumpf B, Schwarzer A, Kalinkowski J, Pühler A, Eggeling L, Sahm H (1991) J Bacteriol 173: 4510
55. Lee W-NP (1993) J Biol Chem 268: 25522
56. Domach MM, Leung SK, Cahn RE, Cocks CG, Shuler ML (1984) Biotech Bioeng 26: 203
57. Wright BE, Butler MH, Albe KR (1992) J Biol Chem 267: 3101
58. Stephanopoulos G, Vallino JJ Science 252: 1675
59. Wiechert W, de Graaf AA, Marx A (1995) in: [60]
60. Schuegerl K, Munack A (ed) (1995) 3rd IFAC Symposium on Modelling and Control of Biotechnical Processes. Pergamon Press
61. Segel IH (1975) Enzyme Kinetics. Wiley, New York
62. McIntyre LM, Thorburn DR, Bubb WA, Kuchel PW (1989) Eur J Biochem 180: 399
63. Varma A, Palsson BO (1994) Appl Env Microbiol 60: 3724
64. Holms WH (1986) Curr Topics Cell Regul 28: 69
65. Vallino JJ (1991) Identification of Branch-Point Restrictions in Microbial Metabolism through Metabolic Flux Analysis and local Network Perturbations. PhD thesis, Massachusetts Institute of Technology
66. Neidhardt FC, Ingraham JL, Schaechter M (1990) Physiology of the Bacterial Cell Sinauer Associates, Sunderland
67. Roels JA (1983) Energetics and Kinetics in Biotechnology. Elsevier Biomedical Press, Amsterdam
68. Barrett GC, Davies JS (1985) in: Barrett GC, Davies JS (ed) Chemistry and Biochemistry of the Amino Acids Chapman and Hall, London, p 525
69. Jones JG, Sherry AD, Jeffrey FMH, Storey CJ, Malloy CR (1993) Biochemistry 32: 12240
70. den Hollander JA, Ugurbil K, Brown TR, Bednar M, Redfield C, Shulman RG (1986) Biochemistry 25: 203
71. de Graaf AA, Wittig RM, Probst U, Strohhäcker J, Schoberth SM, Sahm H (1992) J Magn Res 98: 654
72. Hartbrich A (1995) verfahrenstechnische Charakterisierung von Zyklonreaktoren in der Biotechnologie. PhD thesis, RWTH Aachen
73. Vanni Shanks J, Bailey E (1988) Biotech Bioeng 32: 1138
74. Dori, Arato, Maga (1988) Ann di Chim 78: 529
75. Ni F, Scheraga HA (1989) J Magn Res 82: 413
76. Massiot D, Thiele H, Germanus A (1994) Bruker Report 2: 43
77. Wittig R, Möllney M, Wiechert W, de Graaf AA (1995) in: [60]
78. Ekiel I, Smith ICP, Sprott GD (1983) J Bacteriol 156: 316
79. Eisenreich W, Strauss G, Werz U, Fuchs G, Bacher A (1993) Eur J Biochem 215: 619
80. Strauss G, Eisenreich W, Bacher A, Fuchs G (1992) Eur J Biochem 205: 853
81. Pickett MW, Williamson MP, Kelly DJ (1994) Photosynth Res 41: 75
82. Blum JJ, Stein RB (1982) in: Goldberger RF (ed) Biological Regulation and Development volume 3A. Plenum Press, p 99
83. Anderson DH (1983) Compartmental Modelling and Tracer Kinetics. Springer, New York
84. Lambrecht RM, Rescigno A (1983) Tracer Kinetics and Physiological Modelling. Springer, New York
85. Kuchel PW, Chapman BE (1983) J Theor Biol 105: 569
86. Schuster R, Schuster S, Holzhütter H-G (1992) J Chem Soc Faraday Trans 88: 2837
87. Wiechert W (1995) Metabolische Kohlenstoff-Markierungssysteme – Modellierung, Simulation, Analyse, Datenauswertung. Habilitationsschrift, Universität Bonn
88. Reder C (1988) J Theor Biol 135: 175

89. Wiechert W (1995) in: Dolezal (ed) IFIP TC7 Conf on System Modelling and Optimization. Chapman and Hall, In press
90. Holzhütter HG, Schwendel A (1993) in: [91]
91. Schuster S, Rigoulet M, Ouhabi R, Mazat J-P (ed) (1993) Modern Trends in Biothermokinetics 2, Sept. 23–26, Bordeaux, France Plenum Press
92. Zupke C, Stephanopoulos G (1994) Biotechnol Prog 10: 489
93. Wiechert W, de Graaf AA (1993) in: Bales V (ed) Modelling for improved Bioreactor Performance, Malé Centrum, Publisher & Bookshop, Bratislava
94. Westerhoff HV, van Dam K (1987) Mosaic Nonequilibrium Thermodynamics and Control of Biological Free-Energy Transduction Elsevier, Amsterdam
95. Chambost J-P, Fraenkel DG (1980) J Biol Chem 255: 2867
96. den Hollander JA, Behar KL, Shulman RG (1981) Proc Natl Acad Sci 78: 2693
97. Patnaik R (1992) J Bacteriol 174: 7527
98. Chato Y-P, Liao JC (1994) J Biol Chem 269: 5122
99. Chauvin M-F, Megnin-Chanet F, Martin G, Lhoste J-M, Baverel G (1994) J Biol Chem 269: 26025
100. Mavrovouniotis ML (1993) in: [101] p 275
101. Hunter L, Searls D, Shavlik J (ed) (1993) Proc ISMB-93. AAAI Press
102. Wiechert W (1994) in: Haubensack F, Sühnel J (ed) Bioinformatik, Jena Internet address: ftp.imb-jena.de
103. Fernandez CA, Des Rosiers C (1995) J Biol Chem 270: 10037
104. Schwingenheuer V (1996) Redundanzanalyse bei metabolischen ^{13}C Markierungssystemen. Diploma Thesis, Universität Bonn
105. Holzhütter H-G, Schwendel A, Grune T, Quedenau J, Siems W (1993) Comp Appl Biosc 9: 573
106. Mavrovouniotis ML, Stephanopoulos G, Stephanopulos G (1990) Biotech Bioeng 36: 1119
107. Hofestädt R (1993) in: [101] p 181
108. Wiechert W (1994) in: Gnaiger E, Gellerich FN, Wyss M (ed) What is Controlling Life? Innsbruck University Press (Modern Trends in BioThermoKinetics, vol 3)
109. Hairer E, Norsett SP, Wanner G (1987) Solving Ordinary Differential Equations I. Springer, New York
110. Hairer E, Wanner G (1991) Solving Ordinary Differential Equations II. Springer, New York
111. Cohen M, Bergman RN (1995) Am J Physiol 268: E397
112. Marinov CH, Neittanmäki P (1991) Mathematical Models in Circuit Theory. Kluwer Academic Publishers, Amsterdam
113. Bray D, Lay S (1994) Comp Appl Biosc 10: 471
114. Deuflhard P (1995) Newton Methods for Highly Nonlinear Problems. Academic Press, Oxford
115. Hackbusch W (1993) Iterative Lösung großer schwachbesetzter Gleichungssysteme. Teubner Verlag, Stuttgart
116. Siefke C (1996) Flußschätzung bei metabolischen ^{13}C-Markierungsexperimenten Diploma Thesis, Universität Bonn
117. Seber GAF, Wild CJ (1989) Nonlinear Regression. Wiley, New York
118. Arnold SF (1990) Mathematical Statistics. Prentice Hall
119. Press WH, Flannery BP, Teukolsky SA, Vetterling WT (1988) Numerical Recipes in C. Cambridge University Press, Cambridge
120. Tran-Dinh S, Herve M, Wietzerbin J (1991) Eur J Biochem 201: 715
121. Des Rosiers C, Di Donatos L, Comte B, Laplante A, Marcoux C, David F, Fernandez CA, Brunengraber H (1995) J Biol Chem 270: 10027
122. van Heijden RTJM, Romein B, Heijnen JJ, Hellinga C, Luyben KCAM (1994) Biotech Bioeng 43: 11
123. Wiechert W (1995) In: Schomburg (ed) Bioinformatics: From Nucleic Acids to Cell Metabolism. Verlag Chemie, Weinheim (1995)
124. Bates DM, Watts DG (1988) Nonlinear Regression Analysis and its Applications. Wiley, New York
125. Pázman A (1993) Nonlinear Statistical Models. Kluwer Academic Publishing
126. Walter E (ed) (1987) Identifiability of Parametric Models. Pergamon
127. Delforge J, d'Angio L, Audoly S (1987) in: [126] p 21
128. Cohen SM (1983) J Biol Chem 258: 14294
129. Martin G, Chauvin M-F, Dugelay S, Baverel G (1994) J Biol Chem 269: 26034

130. Portais J-C, Schuster R, Merle M, Canioni P (1993) Eur J Biochem 217: 457
131. Kincaid DT, Pilette R (1992) Comp Appl Biosc 8: 267
132. Meszéna G (1993) In: [91]
133. Reddy VN, Mavrovouniotis ML, Liebman MN in: [101] p 328
134. Gielen G, Sansen W (1991) Symbolic Analysis for Automated Design of Analog Integrated circuits. Kluwer Academic Publishers
135. Cox D, Little J, O'Shea D (1992) Ideals, Varieties and Algorithms. Springer, New York
136. Becker T, Wiespfennig V (1993) Gröbner Bases. Springer, New York
137. Lecourtier Y, Raksanyi A (1987) in: [126] p 75
138. Melenk H, Möller HM, Neun W (1989) Imp Comput Sci Eng 1
139. Posten C, Tibken B (1995) Control Eng Pract 3
140. Tosaka O, Morioka H, Takinami K (1979) Agric Biol Chem 43: 1513
141. Flanigan I, Collins JG, Arora KK, MacLeod JK, Williams JF (1993) Eur J Biochem 213: 477
142. Wiechert W (1995) Comp Appl Biosc 11: 517
143. Mathews CK (1993) J Bacteriol 175:. 6377
144. Sumegi B, Sherry AD, Malloy CR, Srere PA (1993) Biochem 32: 12725

New Concepts for Quantitative Bioprocess Research and Development

Bernhard Sonnleitner

Institute of Biotechnology, ETH Zürich Hönggerberg, CH-8093 Zurich, Switzerland

This paper emphasizes the importance of methodologies and equipment for the credibility of metabolic research. High performance tools for monoseptic cultivation as well as for (non-invasive) on-line analyses are proposed as basis for causal-analytical conclusions and avoidance of artefacts. Special attention is paid to dynamic effects in various relaxation time regimes. A clear and important distinction between macroscopic steady state and physiological steady state is explained and illustrated in examples. Automation and stringent low level process control are shown to be indispensable tools to achieve an ultimate goal: a better quantitative and mechanistic understanding of kinetics and regulation of physiology.

Advances in Biochemical Engineering
Biotechnology, Vol. 54
Managing Editor: T. Scheper
© Springer-Verlag Berlin Heidelberg 1996

1 High Performance Tools

Bioreactors are sterile containments in which biological reactions are performed under well monitored and controlled conditions. They are stepping-stones to achieve bioengineering objectives, namely 1) the preparative production of products and 2) scientific research in many biological disciplines.

The domain of bioengineering is usually regarded as central stage in a production process but it is playing an increasingly important role in the in vivo investigation of living systems (mono- and, hopefully in the future, more and more mixed cultures). The fact that an increasing number of physico-chemical environmental factors can be monitored (quantitated) and controlled opens the gateway for the natural sciences to study complex biological systems including populations of a single one or many distinct species without destruction or disturbance, i.e., in vivo and non-invasive. This new opportunity, made accessible by high tech equipment, permits quantitative studies of biological mechanisms on a higher organizational level than the molecular one, and this is a highly valuable supplementation, not a substitution. The achievable understanding of the regulation of primary and secondary metabolism, sequence control of cell cycle, interspecies signalling or survival strategies of cells is important since it is needed as scientific basis for an economical and ecological optimization of bioprocess operation and control.

1.1 High Performance Bioreactors – (No) Compromises

Biotechnology is an experimental and technical science. Experiments cannot be substituted by model simulations, and material of constant quality must be reliably produced in production runs. Unfortunately, bio(techno)logical processes are regarded as badly reproducible in general. This attribute is seen as an inherent property of biological systems but this historical view is no longer valid. Provided that the processes are conducted in high performance containments which permit near-perfect mixing, afford excellent mass transfer and are tightly controlled with respect to all relevant physical and chemical (micro) environmental properties, one will find those bioprocesses more reproducible than they are said to be. The latter requirement means that variables are actively made parameters which implies that all these variables are permanently monitored and held constant by closed loop control; just the intention to keep several values constant is insufficient.

There is no place for a compromise between quantitative, causal-analytical metabolic research and cheap equipment. Shake flask experiments are good for a first examination. One can derive digital information, e.g., find out whether cells grow or not, whether a product is formed or not. It is eventually also possible to derive some kind of semi-quantitative information, e.g., how rapidly cells might be able to grow in very dilute culture or a rough final yield coefficient (integral method: only stable final values are determined).

It is, however, not possible to derive causal-analytical conclusions because many relevant variables cannot be known, such as the aerobicity of metabolism, transitions into mass transfer limitation, pH etc. It is also not reasonable to derive true rates of conversions since this cultivation equipment is usually not analytically equipped and requires sampling for off-line analyses. This, in turn, requires interruption of shaking and may impose a serious disturbance on the metabolism; this operation is too invasive and is, therefore, not permissive for differential methods (determination of true rates).

The most serious disadvantage of the shake flask system is, however, that calculation of a carbon-balance is principally impossible. Calculation of carbon recovery is the only objective check for the correctness of assumptions and results; regrettably, this fact is ignored in biological research (this becomes immediately obvious if one screens the literature where only very few exceptions can be found). Bioreactors with a higher degree of (automatic) process analyses are mandatory for this purpose and are, therefore, the only acceptable standard for quantitative metabolic research.

There is always place for compromises between the objectives to be met and using more inexpensive equipment. The major objectives of using bioreactors are, besides serving as strictly monoseptic containment, to permit tuning of cultivation conditions as reproducibly and as accurately as possible, to determine as many variables as representatively as possible and, at the same time, to minimize disturbances of the cultures; these are, indeed, subjective qualities. John Villadsen argued [1] that, since it is nowadays possible to determine many state variables such as the biomass concentration or the RNA-content with a precision of better than $\pm 1\%$, it is a matter of our scientific credibility and responsibility to use the appropriate techniques and exploit the available possibilities, anything else would be ignorance.

1.2 Specific Power Input Determines Mixing and Mass Transfer

We definitely need to know quantitatively what a bioreactor's performance is in order to determine what the capability of the biocatalyst is. In other words, one must be able to discriminate between physically and biologically caused effects, such as is the formation of an undesired byproduct the consequence of insufficient oxygen transfer or is it a typical characteristic property of the cells in use? For instance, both *Saccharomyces* and *Candida*-types of yeast form ethanol when cultivated on excess glucose under oxygen limitation (oxygen effect) but only *Saccharomyces*-types form ethanol when oxygen is not limiting (glucose effect) while *Candida*-types do not [2]. Experience teaches that intuition is a bad consultant in these questions and that frequent critical checks of assumptions are necessary (see, e.g. [3]).

At the small scales of laboratory and pilot plant reactors, good mixing and high mass transfer can be easily achieved by increasing the specific power input (P/V) to the reactor. P/V cannot be a topic of concern at this scale because its minimization is not the objective (provided it does not cause cell damage). It is important to be able to set a well defined operating regime which guarantees that mass and heat transfer are not limiting or, if OTR (oxygen transfer rate) needs to be limiting, to know the transfer and consumption rates precisely and have the opportunity to tune them as desired. In order to be prepared it is advisable to characterize the bioreactors in use beforehand (Fig. 1). There is, however, one caveat to be kept in mind: it is not merely the operating conditions such as stirrer speed, aeration rate or filling degree that fully define the transfer capacities, there are also material properties determinant which may change during a cultivation due to excretion and/or (re) consumption of metabolites or medium components. These patters are typical for given biosystems and can certainly influence the volumetric gas transfer coefficient, $k_L a$ (which determines the gas transfer rate), variably.

Mixing is, practically, a lesser problem because many lab scale reactors have a mixing time of ≤ 1 s under reasonable operating conditions [4]. There are virtually no reports identifying such mixing capabilities as limiting factors in bioprocessing, but this cannot be excluded since response times of intracellular key variables such as energy charge to environmental changes in the order of seconds have been documented (e.g. [5] and other contributions in this issue). Immediate inactivation of samples removed from bioreactors is presently the more pressing task than improving mixing.

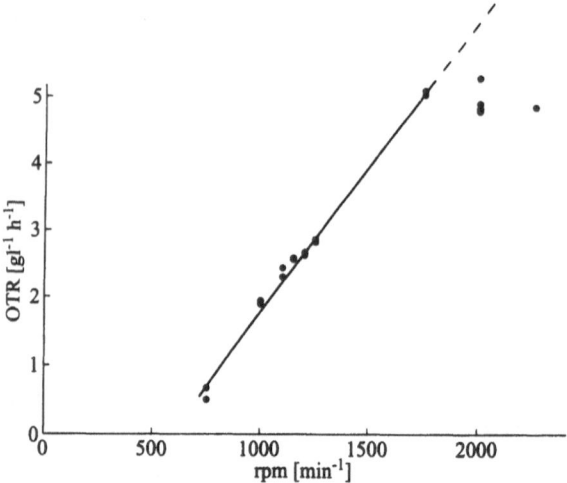

Fig. 1. Oxygen transfer characteristics of a high performance compact loop bioreactor determined using a growing biosystem. Several growth experiments using *Saccharomyces cerevisiae* have been conducted under constant cultivation conditions such as constant temperature, pH, filling degree, aeration, foam breaker speed etc. but stirrer speed (rpm) has been controlled at different values in the individual experiments and oxygen limitation was reached at different relative times: early at low rpm, increasingly later with higher rpm and not at all if rpm exceeded 1800 min⁻¹. During clear oxygen limitation of growth, the oxygen transfer rate (OTR) was determined in each experiment using the pseudo-stationary method: OTR = OUR (= oxygen uptake rate determined from gas balance). There is a minimal stirrer speed (around 600 min⁻¹ in this example) necessary to operate compact loop reactors; below this value, both mixing and dispersion of the gas phase are insufficient

1.3 Foam Control

Foaming biological systems are usually "controlled" by the addition of chemical antifoam agents. This is a common but not an optimal solution to the foam problem because the costs are two-fold: first, one pays for the material to be added – but not as a needed substrate – and, second, the antifoam agent remaining in the harvest stream must be removed from the product during down stream processing. This can become extremely costly as is impressively illustrated by an example from ferredoxin production [Sonnleitner, unpublished]: *Bacillus acidocaldarius* was cultivated in the presence of antifoam agent PPG2000 and a ten-step down stream purification yielded practically no ferredoxin left in "recovered" PPG. After substituting the chemical antifoam with a mechanical foam breaker (Fund-foam), even a three-step purification procedure was sufficient to recover ferredoxin from the cells with more than 95% purity.

tion of medium components into foaming derivatives, degradation of coalescence promoting agents or an increase of free proteins due to cell lysis. It is obvious from these sources that foam formation is not at all constant during a culture, but is highly variable. Over-foaming cultures are useless since the product is lost and the sterile barrier can no longer be kept intact. One can, of course, reduce the working volume of a reactor but this decreases the overall productivity and is disadvantageous. However, foam formation is not a per se-undesirable property of a system because the specific interfacial area of gas bubbles, as, for instance, determined by Lübbert [6] in an *E. coli* cultivation, can be almost doubled due to foam formation. Contrary to non- or sparsely foaming systems, there are no generally useful models available to predict the influence of foam on the gas-liquid transfer capacity or for the distribution of cells between foam and liquid. This is reasonable because a given correlation holds for a specific biosystem in a distinct reactor system and in certain regimes of operating conditions [4]. But that also means that the relations must be re-investigated for any new given culture system.

Mechanical foam breakers are the most valuable, though not optimal, alternative to addition of chemical antifoam agents. The opportunities to compress or destroy foam are manifold, e.g., the use of ejectors, injectors or nozzles, use of rotating disks, lamellae or conical elements, or the use of centrifuges or cyclones. The compaction/destruction of foam (lamellae) derives from either shear, rapid pressure change or, eventually, also from Coriolis force, or a combination thereof. In practice, a foam compression by a factor of two or greater (up to five) is realistic when using mechanical foam breakers.

A comparison of general effects, observed in cultivations run with and without a mechanical foam destructor with various types of foam breakers, shows, according to our experience, that overall oxygen transfer rate (OTR) can be increased by using foam breakers (Fig. 2) and a further improvement is possible by reducing the working volume. This is not so without foam breakers. Volumetric productivity of aerobic organisms, of course, follows the OTR. Foami-

ness of the culture liquid determines the handleability of the system without
foam breaker but it can be significantly extended by foam breakers according to
their performance. Cleaning of systems operated with foam breakers is generally
simpler and faster. Some cells tend to float into foam and accumulate there.
Depending on the drainage characteristics of the foam (i.e., whether it is wet or
dry), the mean residence time of cells in the foam can be long enough to starve
those cells. Samples of liquid removed from the reactor would eventually not be
representative. Immediate compression or even destruction of foam and active
backmixing into the culture liquid diminishes or even eliminates concentration
gradients effectively, which is desirable. Generally, mechanical foam breakers
are very useful tools to improve the credibility of research results (and, of course,
to improve the space-time productivity of a reactor).

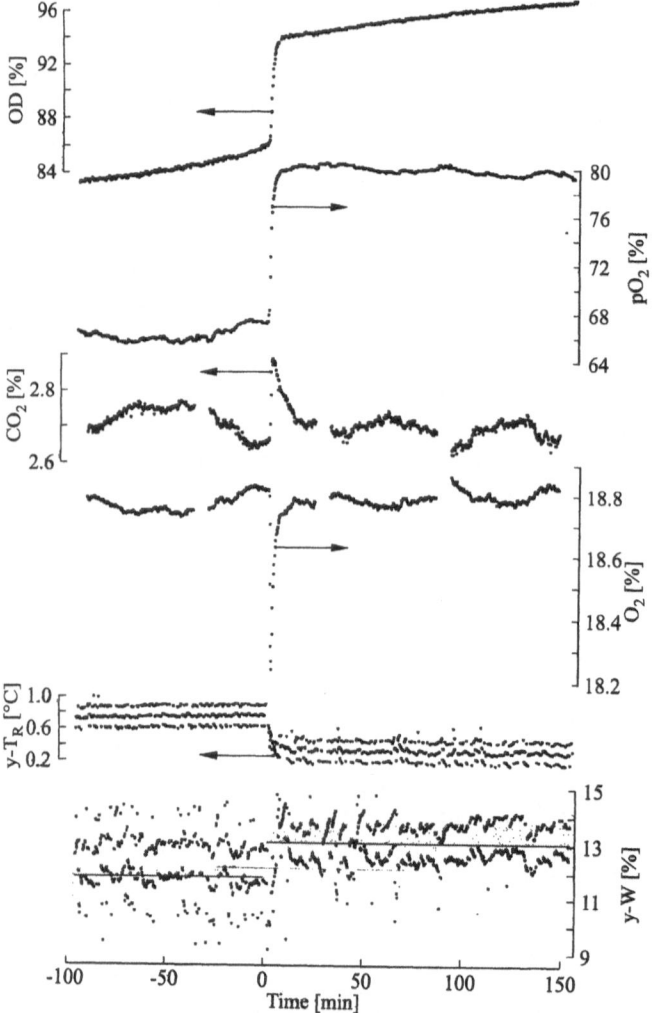

1.4 Membranes Improve Performance

From a bioprocess engineering point of view, process integration is desirable. Whenever a phase separation can be realized in a bioconversion process, one gains another degree of freedom for improving process performance without violating boundary conditions dictated by the biological system itself (e.g., its physiological properties). Some specific advantages are as follows.

1. Volumetric productivity can be increased due to uncoupling of the biological variables μ and q_P – the specific rates of growth and product formation – from the operational variable D – the dilution rate: $\mu = D(1 - R)$, where R is the recirculation ratio (permeate flux/total flux). μ can be tuned to an optimal value, for instance to maximize r_P ($= q_P \cdot x$) whenever q_P is not directly associated with μ. D can be increased above μ_{max} due to cell recirculation – separation of the solid phase – thus boosting up the biomass concentration to $x = Y(s_0 - s)/(1 - R)$ and the productivities Dx and/or Dp.
2. In particular, since D is always $\geq \mu$, the formation of non-growth-associated products is favored when compared to an open production system.
3. Product inhibition can be effectively relaxed if the product is diluted or, even better, selectively removed. In a single-liquid-phase system this is always the case for extracellular water-soluble products in a simple cell recirculation process. In a system with two immiscible liquid phases, the effect can be propagated if the product goes preferentially into either liquid phase which

---

Fig. 2. Secondary effects of the action of a mechanical foam breaker. A continuous culture of *Escherichia coli* grown in a compact loop reactor approaches a new steady state. A critical phase of this transient is zoomed out here: foam formation and relative gas hold-up increase steadily during that period of observation but, at relative time 0, the suspension volume reaches a critical level at which the aerated culture liquid touches the mechanical foam breaker significantly. The foam breaker, looking almost like an over dimensioned Rushton turbine mounted on top of the reactor, starts to act practically as a pump. It transports the foam towards the wall of the reactor from where it is actively sucked back to the bottom of the reactor in the flow regime created by the marine impeller type stirrer. The massive contribution of the foam breaker to power input can be read from the output of the temperature controller (y-T_R: 50% less heating required). This additional driving force increases the liquid flow velocity through the annular space of the compact loop reactor transporting the cells more rapidly. This is partly why the oxygen partial pressure signal (pO$_2$) increases (the probe is mounted at the end of this path just prior to reentering the aeration zone) and, to the dominant part, because an increased gas transfer capacity is achieved as concluded from the exhaust gas composition (O$_2$ and CO$_2$) which is almost constant (except during the event); in this case, the oxygen requirement (i.e., the oxygen uptake rate, OUR) is dictated by the carbon source supply and not by the oxygen transfer rate (OTR): oxygen concentration is definitely not limiting and the system is virtually in equilibrium: OTR = OUR. So the foam breaker also acts as (a good additional) disperger and improves $k_L a$. The consistency of the suspension or foam, however, is clearly changed; this is concluded from the increased optical density (OD; measured with a reflectance probe from Aquasant) and the output of the weight controller (y-W) which drives a peristaltic pump transporting the culture liquid out of the reactor; it needs to transport some 10% more of a foamy liquid with lower mean density, i.e., higher gas content. Although the culture undergoes only smooth, monotonous changes some measurable variables and some operating conditions change abruptly and significantly

can be selectively removed, e.g., by a hydrophobic or a hydrophilic membrane. In any case, biomass concentration and/or product formation rate will increase.

4. Selective recovery of product can be achieved thus shortening and facilitating further down stream processing.

Most of these aspects are predicted to increase relevantly in importance for industrial applications, provided the decisive bottlenecks are identified and practicable methods and solutions are developed [7].

The useful membrane modules are, however, not yet developed to a highly mature state and need much further improvement. For on-line bioprocess analyses, these tools are essential and a fair amount of practical experience is presently available. Membranes are being used for preparation of cell free sample streams or to degas them for on-line analyses using FIA, GC or HPLC, or to achieve phase separation following extraction in FIA systems or to pervaporate gases and volatiles directly into the high vacuum of mass spectrometers [8]. These facts raise important incentives for use on the production scale.

Employing process integration also gives reasonable rise to the expectation that the handling of difficult systems – e.g., heavily foaming ones – will be considerably facilitated. Although technical aspects dominate, now scientific insights in cell $\leftrightarrow$ cell interactions or population dynamics in dense cultures must be expected as well. Some non-invasive instrumental analytical methods – such as NMR-technology [see other contributions in this issue] – will certainly shed some light on the fate of intracellular substances with low relaxation time (e.g., nucleotide phosphates, nicotinamide nucleotides) and the high cell densities required can be achieved in integrated process systems. This will pave the way to completely new fields of research.

1.5 Correct Distinction Between Parameters and Variables

The parameters of a system are those properties which are time-invariant and inherent to the system. In some instances, the experimenter intends to hold one or other property constant by applying a closed-loop controller. The intention is of course necessary but not sufficient for assuring the objective.

Variables are those properties of the system which vary in time, and whose dynamic and steady-state properties are therefore determined by the values of the parameters and other variables. Variables are the levels of biomass (except in a turbidostat, where those too are made parameters [9]) and metabolites, and the fluxes to products such as CO_2, and, of course, other metobolites of interest.

Properties of cultures such as the specific or total oxygen uptake rate and CO_2 evolution rate, though regularly and erroneously referred to as culture parameters, are clearly culture variables. Strictly speaking, they are dependent variables and only time is an independent variable.

Parameters include operational properties such as the pH, temperature or oxygen partial pressure (if controlled), and the composition of the medium which can be known only if it does not contain any complex component (and its feed-rate if in fed-batch or chemostat mode).

Among the parameters are also biological properties such as the genetic and enzymatic make-ups of the inoculum which, generally, are ill-defined [10] and can be kept at a reasonably constant level only when the preparation of any pre-culture follows a very strict SOP (standard operating procedure). It is of paramount importance to include the state and quality of the inoculum in the considerations of cultivation variables/parameters. In seeking to understand the behavior of a biological (or other) system it is, therefore, of determinant importance to distinguish between parameters and variables correctly [11].

2 On-Line Analyses

2.1 Dynamics of Biosystems

The dynamics of microbial cultures in very different relaxation time domains is per se fascinating from a biological point of view, and has important impacts on the characteristics of measurement and process control. The "typical time constant" in a bioprocess is often erroneously anticipated to be equivalent to the entire duration of a cultivation.

Precisely, the relevant relaxation times of a culture system are determined by the actual cell density and the specific conversion rate (capacity) of the culture, i.e., by one or more operational and state variables (for instance feed rate and the concentrations or activities of cell mass and of effectors, if relevant) and inherent characteristic properties of the biosystem which are parameters. There are metabolites with a long lifetime and other (key) metabolites with very short lifetimes, e.g., molecules representing the energy currency of cells such as ATP and other nucleotides. Rizzi et al. [5] have shown that the energy charge response to a pulse challenge of a yeast culture is a matter of a few seconds only. Two other paradigms show that the band width of relaxation times is extremely broad: the time required to achieve a new steady state in a chemostat culture is determined approximately by $(\mu_{max} - D)^{-1}$. The time required by a culture to consume a considerable fraction of small amounts of residual substrate during sampling – thus systematically falsifying the analytical results if no appropriate inactivation takes place – depends, among others, on the cell density and can also be of the order of a few seconds only (Fig. 3).

Almost all models describing the physiology of microbes contain only static kinetic expressions (e.g., the Monod or the Blackman model or extensions thereof) inserted into the differential equations reflecting the mass balances. Thereby, dynamic and pseudo-stationary situations can be described as an immediate response to changes of the microenvironment but the (recent) history

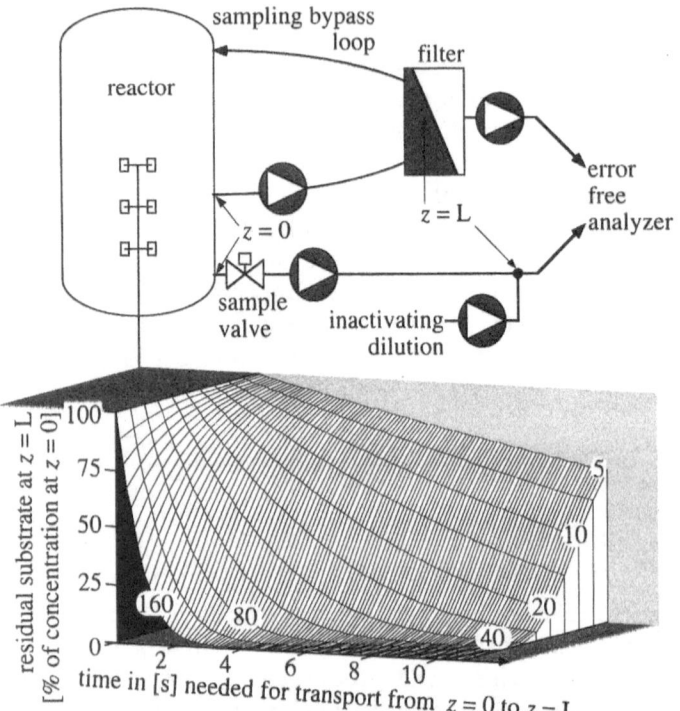

Fig. 3. Origin of systematic errors in spite of potentially error-free analysis. On-line sampling set-ups (*top*) and time trajectories of limiting substrate concentration during sample preparation in the two paradigmatic set-ups depending on the actual culture density (*bottom*). Either is a filter in bypass loop used for the preparation of cell-free supernatant (*upper part in top insert*) or an aliquot of the entire culture is removed using an automatic sampler valve and a sample bus for further inactivation and transport of the samples taken (*lower part*). Both methods require some finite time for sample transportation from the reactor outlet (at $z = 0$) to the location where separation of cells from supernatant or inactivation by adding appropriate inactivators (at $z = L$) takes place during which cells do not stop consuming further substrate. A low substrate concentration in the reactor (namely $s = K_s$) and a maximal specific substrate consumption rate of $3 \, g \, g^{-1} \, h^{-1}$ were assumed in the simulation example to reflect the situation of either a fed-batch or a continuous culture of an industrially relevant organism such as yeast. The actual culture density (in $g \, l^{-1}$) marks some trajectories in the mesh-plot. Please note that the time scale is in seconds

of an actively growing or producing system is ignored. This is why these static models are likely to fail when employed to predict in detail the transient responses to challenges of cultures (rather than to exposure to extracellular limitations).

Nevertheless, such models are very valuable for compensation of systematic errors built into automated analytical procedures. One important example is when sampling requires a well-known but non-negligible time, e.g., when performed using a filter separator operated in bypass or using a sample bus system. A bypass behaves as a plug flow type reactor fraction where flow-dependent spatial gradients develop and where no inactivation must take place because the bulk of the bypassed aliquots is returned to the reactor. The cells continue to consume substrate while they are transported from the exit of the reactor to the filter. The permeate recovered there is representative for the filter site but not for the reactor. Knowing the transport time and some basic kinetic parameters one can easily compensate on-line for such errors provided that there is a useful estimate of the actual biomass concentration available. Even though a bypass can be tuned to be operated at a mean residence time of 5 s or less, this can be enough for significant decrease of substrate concentration in high density cultures (see Fig. 3). Sample buses need a minimal (dead) time for transportation of the sample and in-situ filters tend to fail in high density cultures due to rapid fouling. Hence the problem is real and must not be ignored.

2.2 Continuous Signals and Frequency of Discrete Analyses

Sensors based on either electrical, optical or electro-magnetical principles normally deliver a continuous signal which is very useful. The dynamics of the analyzed system can be resolved according to the time constants of the respective electronic equipment.

In the general case, however, those data will nowadays be digitized but the information loss can normally be neglected because 12 or 16 bit converters are state of performance today.

Furthermore, data are stored in a computer in distinct time intervals resulting in discretization with respect to time and concomitantly in a possible loss of information (e.g. Fig. 4). A data reduction algorithm should therefore be applied which must take account of this fact: raw data should be scanned with high frequency and the essential data – i.e., only when important changes took place – may be stored with the necessary frequency, i.e., variably or not equidistant with respect to time.

Analytical instruments based on either physico-chemical separation methods or relying on (bio) chemical reactions require a finite time to run; these instruments are usually operated in repetitive, non-overlapping batch mode and deliver results with a certain non-negligible dead time; generally, data density is low, for instance of the order of $1 \, min^{-1}$ for an FIA or $2 \, h^{-1}$ for an HPLC.

It is a mere question of the dynamics of the biosystem under investigation whether time-discrete analyses will suffice to describe the behavior with reasonable resolution or not. Time-discrete signals are, especially if only available at a low frequency, difficult to use in closed loop control. However, there are solutions to this general problem provided the variable of interest can be continuously estimated (directly or indirectly). For instance, repetitive calibration of the continuous signal for partial pressure of a volatile component generated by a membrane interfaced MS-probe (i.e., the estimate) with the time-discrete concentration signal produced with an on-line GC (i.e., the measurement) can effectively compensate for drifts or alterations of the membrane permeation behavior and renders the estimate suitable for process control (for details on the example acetoin see [12]).

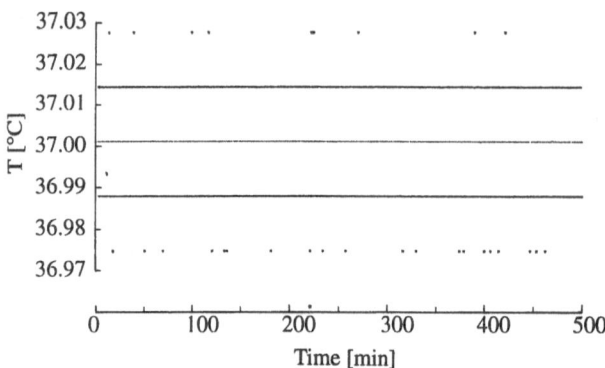

Fig. 4. Example for discretization of continuous signals. A high resolution plot of reactor temperature, which is continuously measured with a Pt-100 element, shows 1) a perfect temperature control, and 2) the two types of discretization that take place on a digitally controlled equipment: the 12-bit conversion of the measuring range used (0–50 °C) discretizes the signal in approximately 12 mK steps. Only slight deviation from the setpoint (37 °C) is needed to drive the controller. Only the first, two intermediate (possibly outliers) and the very last data point would be written to a documentation file when the data reduction algorithm were applied to this set of data using a window-width of (usually) 50 mK

2.3 Systematic Analytical Errors

Sampling and sample preparation usually cause systematic errors in the results of analytical procedures. It is, therefore, important to be able to reproduce these errors and to know them as accurately as possible, as discussed above, in order to compensate for them by appropriate calculations. Good sampling techniques such as rapid automatic sampling, fast cooling, heating or immediate chemical inactivation help to minimize these errors. But there are other errors as well. Some are associated with interferences of a selected method to unknown compounds in the analyte (i.e, in the non-analyzed matrix containing the component of interest). We have found that the results from glucose analyses can depend on the detection method: for instance, in undiluted mode, there are generally higher readings to be expected from a Yellow Springs analyzer which employs an amperometric H_2O_2-electrode than from a colorimetric detection (using ABTS and peroxidase) in spite of the fact that the sample pretreatment is identical, namely removal of cells by using a by-pass filter within 2 to 3 s (unpublished). These discrepancies usually vanish when the samples are diluted. Although the manual of the Yellow Springs analyzer contains a long table of known interfering substances from which a potential systematic error could be estimated, the composition of the analyte is normally not known (both qualitatively and quantitatively) and, hence, the table is often meaningless in practice. The interferences do not affect the reproducibility under otherwise constant conditions but they can severely affect the accuracy of results. This difference is important in this context. Our procedure to solve this problem has therefore developed to the following.

1. Compare the results obtained from identical samples with different methods.
2. If there is a significant deviation, dilute the samples (more) and repeat. It the deviations vanish, accept these values.
3. If not, spike the analyte with an accurately known amount of glucose and estimate the probable individual errors from the (precisely known) added amount and the analytically recovered surplus.
4. Consider mutarotational non-equilibrium (in case of glucose analysis, as discussed below).

There is yet another pitfall associated with some glucose analyses, specifically with those methods using glucose oxidase. Such methods are very common since the enzyme is relatively inexpensive and very stable. This enzyme is known to react with β-D-glucose but not with α-D-glucose [Merck Index]. Solid glucose (monohydrate) exists in the pure α-form when crystallized from water below 50 °C and in the β-form when prepared from hot water, dilute acid or pyridine. The specificity of glucose oxidase has important analytical consequences as first pin-pointed by [13].

In aqueous solution, a mutarotational equilibrium of the two anomers is reached spontaneously – but not instantaneously – in which the ratio of $\alpha : \beta$ is 36:64 at a temperature around 30 °C. The enzyme mutatrotase accelerates mutarotation considerably. The rate at which the equilibrium is reached spontaneously depends greatly on pH, temperature and other solution components. Benthin et al. [13] used numerical values for the time constant of mutarotation (k_{mut}) in the order of 2.8 to 11.4 h^{-1}. We have determined k_{mut}-value = $1.2 \pm 0.02\ h^{-1}$ in pure water and $12 \pm 1\ h^{-1}$ in phosphate buffer as well as in defined E. coli growth medium independent of whether starting from pure α- or β-form [14]. These numerical values of the time constants confirm that there is no time for significant mutarotation to take place in rapid on-line analyses. If glucose uptake is anomerically specific, as shown to occur [13], analytical results obtained with a glucose oxidase method must be corrected accordingly. The same holds true for analyses of rapid biological transients and for the above cited spiking method (if the spiking solution happens to be freshly prepared). How dramatic the consequences on research results can be is demonstrated in the paradigm discussed in Fig. 6 later.

A further important example of systematic errors is analysis of exhaust gas composition and calculation of mass balances therefrom. All analyzers create a signal which is proportional to the number of molecules in a given volume (that of the measuring chamber), and this is not directly a measure for the molar fraction of a component in the gas; specifically, the signals depend on the pressure in the measuring chamber. With a mass spectrometer at hand, one can easily determine an inert gas component (e.g., N_2 or Ar) as well as the reactants of interest (mostly O_2 and CO_2). One can, then, normalize these signals to the inert gas signal. Using an inert gas balance, one gets rid of the dependence on pressure, e.g.,

$$RQ = \frac{\left(\dfrac{m_{44}}{m_{\text{inert gas}}}\right)^{\text{out}} - \left(\dfrac{m_{44}}{m_{\text{inert gas}}}\right)^{\text{in}}}{\left(\dfrac{m_{32}}{m_{\text{inert gas}}}\right)^{\text{in}} - \left(\dfrac{m_{32}}{m_{\text{inert gas}}}\right)^{\text{out}}}$$

where m_i is the signal intensity measured at a mass/charge ratio $= i$ (which is 44 for CO_2, 32 for O_2 and eventually 40 for Ar) under the assumption that m_i depends linearly on the molar concentration of component i in the measuring chamber (to be confirmed by calibration). When a classical analyzer system is used, the direct information of inert components is not available and the molar fraction of the inert gas must be calculated as $(1 - yCO_2 - yO_2)$. Then, a formula such as

$$RQ = \frac{y_{CO_2}^{\text{out}} - y_{CO_2}^{\text{in}} - y_{O_2}^{\text{in}} y_{CO_2}^{\text{out}} + y_{O_2}^{\text{out}} y_{CO_2}^{\text{in}}}{y_{O_2}^{\text{in}} - y_{O_2}^{\text{out}} - y_{O_2}^{\text{in}} y_{CO_2}^{\text{out}} + y_{O_2}^{\text{out}} y_{CO_2}^{\text{in}}}$$

can be used to derive desired dependent variables. However, if the molar fractions for the reactands, yCO_2 and yO_2, are error-biased, then so is the calculated value for inert gas, y_{inert}, error-biased and the final result as well. Heinzle et al. [15] have analyzed the error propagation in detail with respect to model based biomass prediction and Locher et al. [16] have shown the influence of slight calibration errors or drifts on the calculation of a software sensor (the biological variable RQ). In the context of mass balancing, where the correctness of the absolute values is decisive, the appropriate consideration of pressure on the measurement result is important. This can be done by keeping the pressure in the measuring chamber constant using an automatic pressure regulator, by measuring this pressure separately, or by frequent recalibration (e.g. once per hour).

Even error-free determination of CO_2 in exhaust gas does not necessarily result in correct determination of CO_2 production (rate) or carbon recovery if the solubility and reactivity of CO_2 at neutral or alkaline pH is not taken into consideration. CO_2 dissolves in water to hydrated CO_2, $H_2O.CO_2$, which is in equilibrium with the unstable H_2CO_3. This, in turn, is in practically instantaneous equilibrium with HCO_3^-, determined by the pH of the solution

$$CO_2 + H_2O \underset{k_2}{\overset{k_1}{\rightleftharpoons}} H_2CO_3 \xleftrightarrow{\text{spontaneous}} HCO_3^- + H^+$$

with the kinetics

$$r_{c>b} = k_1\, c - k_2\, [H_2CO_3] = k_1 c - k_2 \frac{b\, 10^{-pH}}{K_b}$$

where $r_{c>b}$ is the net reaction rate, c is the concentration of dissolved CO_2, b is the concentration of bicarbonate (HCO_3^-) and K_b is the dissociation parameter of carbonic acid to bicarbonate. k_1 and k_2 are rate parameters for which quite different sets of numerical values are found in the literature. This most probably has to do with the different acidity of $H_2O.CO_2$ and H_2CO_3. The latter is the stronger acid. A comparison of the two parameter sets shown resulted in faint differences only: Royce [17] reported $k_1 = 209 \, h^{-1}$, $k_2 = 2160 \, h^{-1}$ and $K_b = 6.8 \, 10^{-4} \, mol \, l^{-1}$ and Noorman et al. [18] used $k_1 = k_2 = 134 \, h^{-1}$ with a $K_b = 4.7 \, 10^{-7} \, mol \, l^{-1}$. The Handbook of Chemistry & Physics reports a value of $K_b = 4.26 \, 10^{-7} \, mol \, l^{-1}$.

Anyhow, the important point is the conclusion that a significant amount of CO_2 may not be found in the exhaust gas but rather in the culture liquid in the form of bicarbonate or even carbonate (in alkalophilic cultures). For example, Ponti [19] estimated that up to 25% of total carbon mineralized in sewage sludge treatment were captured as bicarbonate and not released via the gas phase. Disregarding this fact must result in systematically erroneous determinations of respiratory quotient (RQ), carbon dioxide production rate (CPR) and carbon recovery.

3 Control of Bioprocesses

3.1 Estimation and Control of Physiological State

The estimation of the physiological state of a culture involves more than one (measurable) variable at a time, i.e., recognition of complex patterns. Various algorithms have been used for this purpose [20–25]. All of these have in common the fact that it is not the present values alone that are evaluated, and there is always the recent history of a set of individual signal trajectories involved.

In some cases, the data describing the actual state and their recent history are compared with so-called reference patterns: these are data from historical experiments or runs which an expert has associated with a "typical" physiological state. A physiological state is recognized either if the actual constellation matches any one of the reference set best – in that case, there is always an identification made – or if the match exceeds a predefined degree of certainty, e.g., 60% – there it can happen that no identification or association is possible with too high a limit selected. The direct association with reference data needs normalization (amplitude scaling) and, eventually, frequency analysis in order to eliminate dependencies on (time) shifts, biases or drifts.

In other cases, the data trajectories are translated into trend-qualities via shape descriptors such as glucose uptake rate is decreasing (convex down) while RQ is increasing (concave up). These combinations of trends of the trajectories of various state variables or derived variables define a certain physiological state; the advantage of this definition is that the association is no longer dependent on time and on the actual numerical values of variables and rates.

3.2 Selection of an Operating Point on the Safe Side

The major impacts for optimization ($\equiv$ effect) of biotransformation process – i.e., of the controlled proliferation of cells on distinct substrates and the formation of products – come from the identification ($\equiv$ basis) of determinant key elements – the decisive structures, functions, rates and regulatory mechanisms.

Regulatory mechanisms have their roots in molecular mechanisms. The function of biologically active molecules or aggregates thereof is determined by their molecular structure. The regulatory mechanisms on the level of populations (e.g., growth, di- or polyauxia, product formation, synchronization) are in fact a superposition of many individual but interdependent molecular mechanisms. Therefore, they cannot be studied exhaustively and exclusively at the level of isolated biological subsystems (e.g., enzymes, ribosomes, membranes) because important components are removed by isolation. Rather, an ideal experimental system is an undisturbed and well controlled culture of cells. Non-invasive process analyses are to be preferred in order to exclude artefactual conclusions from the responses of the biological system to any alterations of the microenvironment caused by the analytical method.

Because of the complexity and multiplicity of the components of a living system there are simplifications necessary, for instance, reduction to cause → effect mechanisms. This means that dominating key variables like concentrations, activities or fluxes and functional entities must be identified as relevant, and, if they are, quantified. Structured approaches will permit descriptions of the dynamics of regulation.

Some of the relevant variables are obvious but many are still obscure. However, the exhaustive perception of them is determinant for the success of a process because it is a prerequisite to pin-point the so-called culture parameters (i.e., variables made constant by means of closed loop control) correctly and efficiently. Whenever the chain of effects determining the physiological state is well known, there is normally a functional control strategy and a realistic way available to force a culture into the desired state. Only then can the real needs for process analyses be deduced conclusively.

Unfortunately, this is nowadays not the case for most academic and many technical bioprocesses and poses serious problems to all. That is why so many processes are operated off the optimum – on a safe side, for instance, at a feed (or dilution) rate lower than the one allowing maximal productivity – but the solution of the future must be via causal-analytical investigations and appropriate process-technological consequences.

Analyses of strictly monoseptic bioprocesses are comparatively underdeveloped. Application of highly sensitive and selective sensors for important components in bioprocesses is certainly not routine. The backlog demand in bioprocess monitoring, as compared to medical diagnosis, has its roots in the sterile barrier of the technical process (which has no immune system like humans), in the need for a high dynamic range of the measured variables and in the varying interferences with the analytical matrix. There has been a recent answer to the first and third points, but the second is not yet solved. A new glucose biosensor, which can be sterilized using ethylene oxide, has been successfully applied in medical analysis but there the dynamic range is very narrow [26].

3.3 Definite Knowledge of Limitations

Among the operating conditions of a cultivation or biotransformation there are such that can be easily measured and tuned (like temperature, pH, aeration or agitation rate) but there are equally important properties that are too easily ignored such as the composition of the medium or the quality of the preculture. Causal-analytical conclusions are only reasonable if the real causes can be identified. And this is exactly what is likely to be very difficult or even impossible when complex media are employed and precultures are neither monitored nor controlled with respect to the most basic "culture parameters". We need to know the limitations of growth and/or product formation precisely at any time in (scientific) metabolic research.

A growth limitation imposed by a medium component can be positively identified by analyzing the transient response of a culture to pulsed addition of the respective substance. This is ideally done in continuous culture because of simplest interpretation but it is not restricted to this mode of operation. A pulsed addition of the hitherto limiting substrate relaxes the limitation thus raising the specific growth rate towards its maximal value for some time. Cell density increases in continuous culture because $(\mu\text{-}D)$ becomes positive. The concentrations of all other essential medium components must decrease due to the additional consumption caused by the increasing biomass. This may then lead to an intermediate limitation of growth by another component which had been provided in the medium in only little excess. After the concentration of the pulsed component has returned to the original steady state level the additional consumption of this component continues for a while because the sink term in the substrate balance $-q_s x$ (with $q_s < 0$) – is greater than in steady state: the biomass concentration is "too" high. This results in a decrease of μ below D and cells start to wash out. Concomitantly, the oversized sink term shrinks allowing the substrate concentration and the specific growth rate to approach the steady state value again. Figure 5 shows the respective time trajectories provided the cells are able to respond immediately to the extra substrate supply.

Cells, however, can also suffer from an intracellular limitation. Cells grown for a long time at low specific growth rate imposed by limited substrate supply (for instance, at low dilution rate) will, for economic reasons, not produce maximal amounts of enzymes necessary to cope with a suddenly increased substrate

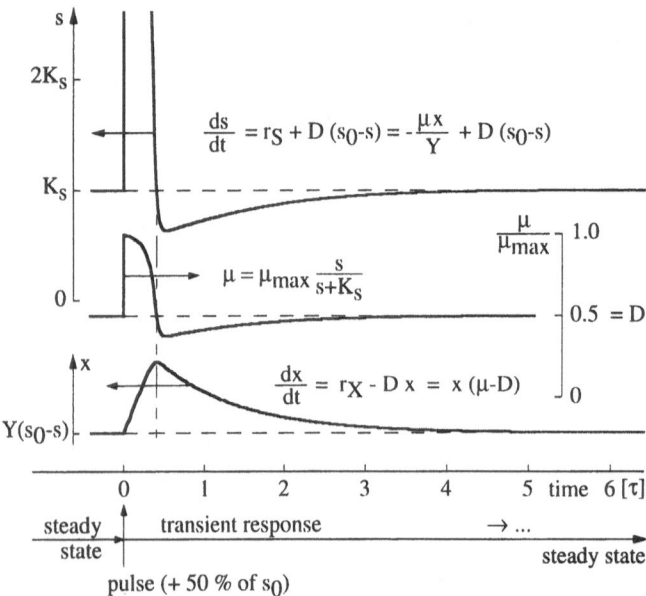

$$\frac{ds}{dt} = r_S + D\,(s_0\text{-}s) = -\frac{\mu x}{Y} + D\,(s_0\text{-}s)$$

$$\mu = \mu_{max}\,\frac{s}{s+K_s}$$

$$\frac{dx}{dt} = r_X - D\,x = x\,(\mu\text{-}D)$$

Fig. 5. Responses to a pulsed addition of the limiting substrate. The example assumes a continuous culture operated at dilution rate $D = 0.5\,\mu_{max}$. *Top*: time trajectory of the limiting substrate (s) at $s = K_s$ prior to the (Delta-Dirac-) pulse which increased the concentration to 50% of continuous feed (s_0). *Middle*: trajectory of specific growth rate μ. *Bottom*: trajectory of biomass concentration (x, which has a steady state value of $Y(s_0 - s)$). Note that both substrate concentration and specific growth rate under-shoot their steady state values. The inserted balance equations are valid except during the application of the pulse at $t = 0$

Fig. 6a, b. Macroscopic and physiological steady states are not identical. Dynamic responses of *Saccharomyces cerevisiae* to shifts of dilution rate have been analyzed after the culture had been operated undisturbed for approximately 10 mean residence times at $D = 0.05\ h^{-1}$: the symbols refer to FIA-measurements of glucose (using glucose oxidase which is specific for ß-D-glucose; however, calibration was made with glucose solutions in mutarotational equilibrium and the medium fed was also in mutarotational equilibrium). The response to the first shift-up of D (at $t = 0$) shows clearly an overshoot of glucose, the second does not and the third is not significant enough: **a** using a model which takes the uptake-selectivity for α-D-glucose of *Saccharomyces cerevisiae* according to [13] into account allows one to analyze the experimental observations. The model further assumes that the enzyme set responsible for glucose uptake and/or catabolism responds dynamically to the supply (as described for the respiratory enzymes by [3]). At $t < 0$, the cells had approximately 2.5% of the full enzyme set available which is equivalent to the necessary fraction of maximal glucose consumption rate at this dilution rate ($D = 0.05\ h^{-1}$). While the increase of D (*lower panel*) is stepwise the response to the increased feed is a delayed production of enzymes thus resulting in an intermediate shortage of enzymes (*middle panel*) and, therefore, an overshoot of glucose. The measurable overshoot is strong because the less selectively consumed anomer ß-D-glucose is analyzed species. However, the model shows the α-anomer and the sum of both ('glucose') as well (*top panel*). After reduction of D, the enzyme set is not assumed to be actively degraded but rather to be diluted due to growth. The enzyme set is still sufficiently large when the second shift-up occurs so that no overshoot of glucose is caused whereas the simulation predicts for the third shift-up a very small overshoot, the amplitude of which is within the noise of measurement and cannot be verified experimentally; **b** the model prediction for a highly expressed enzyme set ($\gg 5\%$) which corresponds to a culture grown at a dilution rate greater than $0.05\ h^{-1}$ most recently: no overshoot is to be expected at any time in such a situation (contrary to experimental findings). Although macroscopically in steady state prior to the pulse, this (simulated) culture had not yet attained its physiological steady state

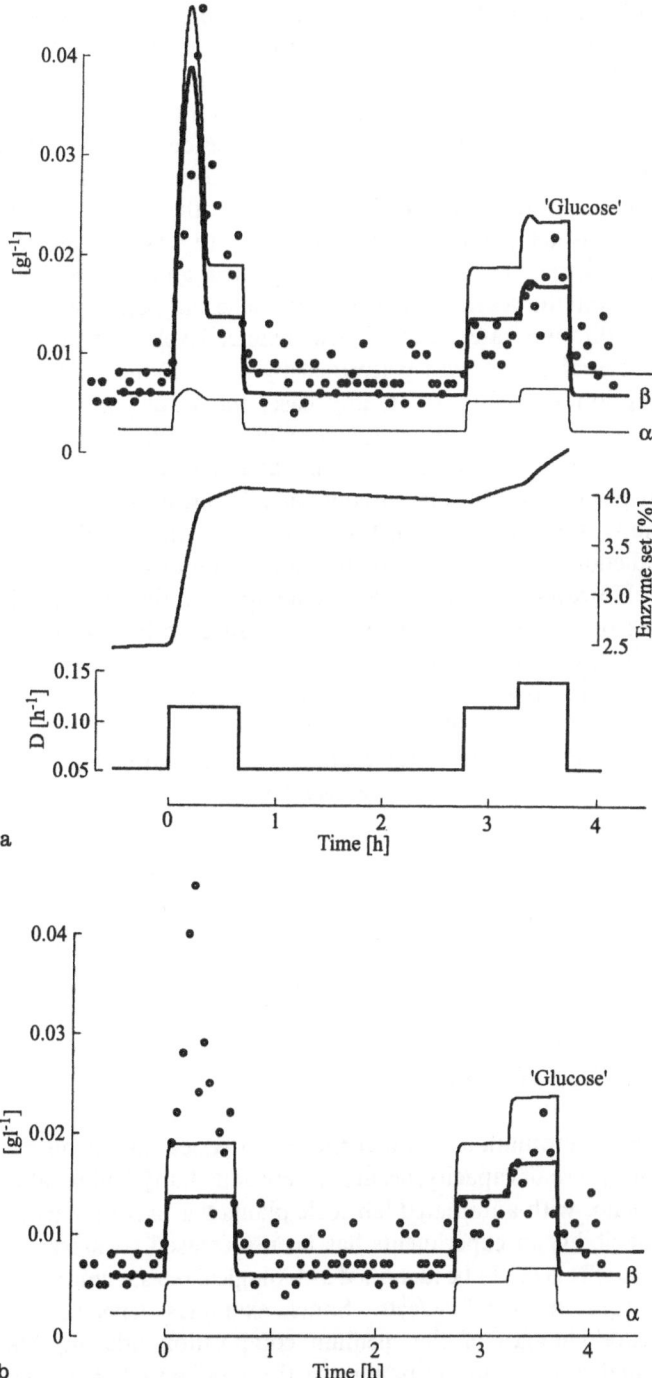

supply. This can also be concluded from the fact that the RNA content of slowly growing cells is significantly smaller than that of cells consuming substrate rapidly and therefore also growing rapidly [27, 28]. We have followed this aspect experimental with *S. cerevisiae* and *E. coli*.

The starting point for dynamic experiments was always a culture grown at low dilution rate (10–20% of μ_{max}) for at least five mean residence times. Then the cultures have been challenged by an increase of substrate supply (shift-up of dilution rate) for a limited time of observation after which the original conditions have been re-established. After some time, sufficient to allow the macroscopic steady state to settle, the challenge was repeated and the response analyzed. An example from the yeast system shown in Fig. 6 was performed with a stepwise change of dilution rate between two distinct levels and another example from the bacterial system was performed by a non-monotonic rampwise change of dilution rate (Fig. 7). In any case, the results from both biosystems confirm the idea of an economic intracellular household and illustrate the need to differentiate between macroscopic and cellular (or physiological) steady state. In other words, the macroscopic steady state is dictated by an extracellular limitation – the substrate is just not available – and the physiological steady state is dictated by the intracellular composition or activity of the enzymic apparatus. However, the causes are different and the effects (i.e., measurable responses of a biosystem) can be distinguished with appropriate equipment in a suitable experimental design.

An important, yet speculative, conclusion is that oscillatory operation of a bioprocess might render it more stable and more productive since variations of operating variables are tolerated and compensated by a repeatedly challenged biosystem (which is less fit in steady state; see also 5.3).

4 Automation

4.1 Reliability and Reproducibility

There is no way around automation in biotechnology. It does not encourage indolence, it serves safety, use of capacity, accuracy, reproducibility and reliability exclusively. Experience with automated lab scale plants has exceeded many expectations. Reproducibility of experiments has been increased so markedly that many macroscopic effects could be identified as biological ones, quantitated and discriminated from technical effects. Some examples (specified in [3, 16, 29–31]) are trace changes of the medium composition affecting the byproduct excretion and reconsumption pattern or the exponential characteristic of growth significantly, minute changes (< 1% off the intended operation

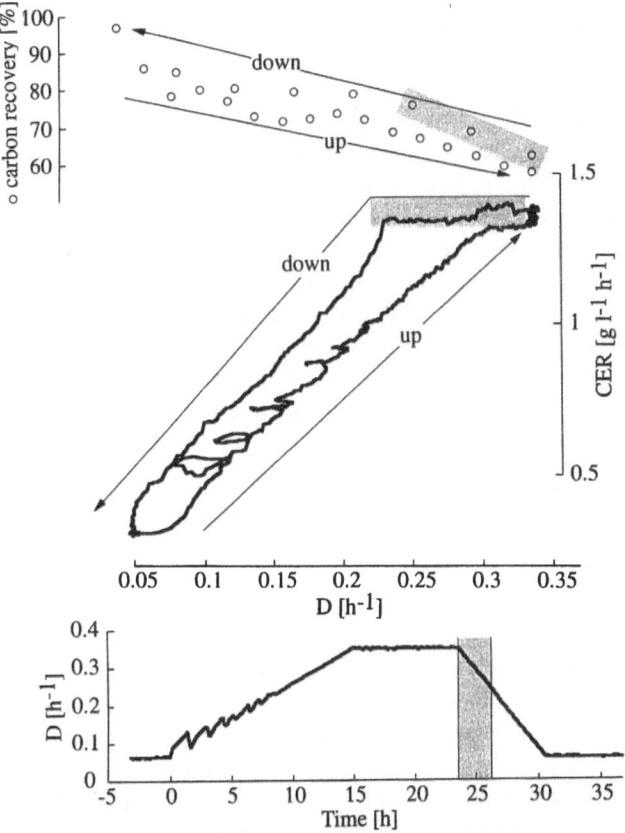

Fig. 7. Intra- and extracellular limitation regime in *Escherichia coli*. A continuous culture was operated for more than five mean residence times at low dilution rate (0.05 h^{-1}) and then challenged by a ramp-wise shift-up of dilution rate, held at $D = 0.35$ h^{-1}, and shifted down again: carbon recovery based on residual substrate, biomass and CO_2 was calculated (*top panel*) and the carbon dioxide evolution rate (CER; *middle panel*) was determined. The flow-rate controller was badly tuned during the initial shift-up phase resulting in an oscillatory superposition on the linearly increasing set-point (*bottom panel*). The cells obviously do not have sufficient intracellular resources (e.g., enzymes) available during the shift-up which results in the excretion of an overflow metabolite (acetate data not shown and others not analyzed) which can be reconsumed whenever the feed decreases. The total conversion rate (of fresh substrate fed and reconsumed overflow metabolite) seems to remain fairly constant since CER remains almost constant even though D decreases (*top parts of loops of CER vs D in the branch marked "up"*). Probably, the intracellular enzyme set is only increased when cells are challenged with more glucose and is, of course, delayed. When the operating conditions are switched to extracellular limitation (namely less supply during shift-down), the cells are getting more and more capacity to reconsume overflow metabolites (*grey shaded areas*). After some time, the cells have an intracellular over-capacity: all substrate is most effectively converted into cell material and no carbon containing overflow metabolites are produced any longer; the oxidized carbon appears only in the form of CO_2. The CER reaches maximal values at any given D during the *branch marked "down"*, i.e., when there is no intracellular limitation only. Data from [81]

point) of feed rate affecting period and amplitude of synchronous oscillations extensively, or faint transitions into double-limitation regimes (e.g., glucose and oxygen) affecting the product formation greatly.

We have seen indications that the sensitivity of responses of a biological system could be even better than that from the presently available direct measurement methods and/or devices for physical variables, i.e., biological responses could be exploited as very sensitive sensors. There is one example for dilution rate shown in Fig. 8. We are able to monitor the weight of the reactor with a precision of better than $\pm 0.05\%$ of the working "volume" [which is equivalent to weight provided density does not change] and to determine the feed rate from the weight loss of a buffer reservoir over time with an accuracy of ± 0.1 g (equivalent to 0.05–0.2% of the balanced volume) and 1 s, respectively, resulting in a precision of the dilution rate control of better than $\pm 1\%$ at best (due to controller behavior and error propagation). However, this is not good enough to establish a perfect steady state as is obvious from the CO_2 signal trajectory shown which follows the dilution rate with the dead time of exhaust-CO_2 determination.

The exploitation of these possibilities must be promoted in future.

4.2 Error Reduction

Automation makes more complicated set-ups workable; for instance, small pilot or production plants with several stages or with integrated down steam processing can easily exceed 50 analogue I/Os and have more than 200 digital I/Os. It is definitely not possible for a (single) human operator to monitor and tune these I/Os in real time error-free. A similarly problematic situation is encountered with systems when critical sequences during preparation (e.g., sterilization) and operation (e.g., repetitive (fed-) batch operation) are no longer suitable for human supervision and manual execution.

However automation of bioprocesses does not only mean automatic execution of a priori determined sequences or algorithms, simply operations of actuators handed down to machines; it includes automated analysis of the bioprocesses as well. The data path opened by automation includes real-time analysis of generated data, immediate plausibility checking, calculation of software sensors and eventually the communication with intelligent analytical subsystems. The data archive can be reliably replenished, i.e., loss of documentation can be effectively eliminated.

4.3 Exploitation of Time

Automation also permits exploitation of 24 hours a day and 7 days per week. Automation in bioprocessing also has an important social effect, as boring, routine tasks can much more efficiently and reliably be delegated to machines

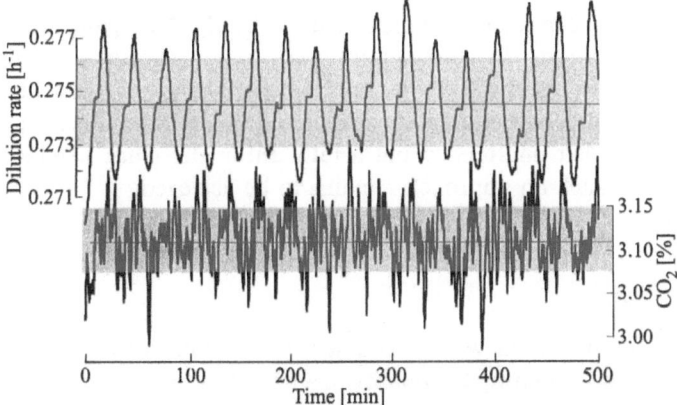

Fig. 8. Biological responses can be very sensitive indicators for inconstant operating conditions or "a blow-up of a so-called steady state". The dilution rate (D) in a glucose limited continuous culture of bakers yeast was controlled to ± 1% of the set-point (*top panel*) by better separate controls of feed rate (F) and reactor weight (V; both not shown; D = F/V) and CO_2 in the exhaust gas was measured with 12 bit resolution in the range [0–5%] (*lower panel*). Although noisier than the D-signal (because air flux and pressure were also separately controlled to approximately ± 1% each and influenced the exhaust stream), the low frequency fraction of the CO_2-signal follows the D-signal exactly with the dead time required for gas transport into the analyzer. Mean values are indicated by *horizontal lines* and standard deviations by the *shaded areas*. CO_2 production is really a good indicator for the quality of the dilution rate control. The noise of the CO_2-signal is technically caused and reflects just a biological response

and, hence, personnel can be relieved from these tasks and engaged in more interesting work. It replaces operators for being responsible for all the details to be carried out correctly in real time which is, anyhow, questionable in larger installations. This is sufficient reason to implement some automation in industry in order to improve and guarantee a constant high quality standard of a bioprocessing facility with high precision, and to gain independence of specific personnel. But there are more objectives to be met: automation not only increases the reproducibility of routine production processes, it is also a most effective tool for causal-analytical physiological research and for the development of bioprocesses.

We have applied the chemostat-pulse-and-shift technique automatically for medium development several times. The techniques for applying pulses and shifts are easily automated [32], even the evaluation of the responses is relatively simple provided the responses can be reliably measured on-line ([33]). However, an effective methodology is still missing which would permit automatic planning of experimental design, e.g., selection of substance(s) to be tested next with maximal probability to provide the actually needed information; unfortunately, this is still a matter of subjective decision and deserves further strategic development. There are other new approaches which allow automation, namely evolutionary algorithms, but they need to evaluate "populations" which requires a massive parallel work load rather than a serial one [34]. This has been realized so far only in experiment series using shake flasks; the disadvantages of this technique, however, cannot be compensated for by the time gain due to automated experimentation. In bioreactors, we have successfully used a Simplex-algorithm for optimization of operational conditions in sewage sludge treatment [35]; after defining the starting conditions and the constraints, the algorithm could be left unattended but a fully automatic plant would be needed. Realization of such an approach remains to be achieved.

4.4 Validation

On-line measurements produced with in-situ sensors are difficult to validate. The usual procedure for evaluating the quality of a measurement is restricted to calibration/checking prior and after a cultivation. A few sensors such the pCO_2- or the Cranfield/GBF-glucose sensor [36] allow removal (at least of measuring buffer but also of the transducer itself) and, therefore, recalibration of the transducer during a cultivation.

Chromatographs and FIA-systems can be regularly recalibrated but the sterile interface is not. In these cases, an important part, but not the entire measuring chain, can be validated, and the membrane (or other interface) plays a decisive role in affecting the overall result. Other sensors such as pH- and pO_2-probe can be mounted via a retractable housing which allows either sterile exchange or withdrawal for external recalibration during a run. A further possibility of gaining some information about the reliability of a measure is to mount a number (> 1) of identical sensors in comparable positions and to check the individual signals for equality. This technique has been exploited for many years in other technologies but the inherent pitfall is that systematic errors are validated rather than the correctness of the measured value itself.

It is highly desirable to have alternative principles of measurement at hand which are operative at the same time. For instance, non-inert gases can be estimated using completely different analyzers: both oxygen and carbon dioxide can be determined in exhaust gas using either the classical analyzers sensitive for infra red absorption (for CO_2) and paramagnetic properties (for O_2), respectively, or using a mass spectrometer. Stability of these instruments is reasonable and interferences from other components are not very likely. However, variation of the pressure in the measuring site must be compensated for. The partial pressures of both gases can be monitored with membrane covered probes. The membranes are the weakest elements in this measuring chain and are most difficult to validate experimentally. The electrodes or the mass spectrometer behind the membrane are just transducing elements and quite reliable. The measurements obtained from the gas and from the liquid phase quantify, of course, different state variables but they are tightly linked via gas-liquid mass transfer. A comparison of these data allows at least a reasonable consistency check and identification of malfunctions. With expert knowledge of the gas transfer variations one could even quantify the plausibility of the individual signals. Pattern recognition can help generally to identify malfunctioning sensors in some cases [37]; in other this unfortunately still remains a subjective task [38].

There is a set of various biomass sensors available which is very fortunate for the reasons discussed above. However, not too many useful comparisons and validations have been made in recent years [9, 39–57]. The sensors used in these studies exploit various measurement principles, including optical (reflectance, absorbance, scatter and fluorescence, with classical and laser light sources), electrical (permittivity) and morphological properties (microscopic, image analysis) of cells or cell suspensions quantified. The individual methods have different sensitivity and selectivity; they need very careful calibration to both the biological and reactor systems in use. Although no one completely satisfactory statement can be found in these investigations it would be quite interesting and helpful to cross-validate as many as possible of those alternative methods in identical experimental runs in various bioreactors. The increased information created is expected to result in more reliable quantitation of biomass with a relatively narrow confidence interval.

4.5 Automatic Limitation of Damages, Data Losses and Artefacts

In and around a bioreactor there are many technical installations with moving, sealing and sensing parts such as motors, double mechanical seals, membranes, valves, pumps, sensors, and analytical instruments which all require maintenance and, quite important, some care. A reactor is only useful if almost all parts work properly and reliably at the same time, or it is as useful as its weakest part at any one time. Therefore, it is very reasonable to allocate some essential elements of care-taking tasks down to an automatic system.

Elements essential for safety should also normally be realized in dedicated hardware (e.g., built-in rupture disks or safe position of valves in case of power failures). Besides such established "good practice" methods we have experienced that additional software solutions are really advantageous since they help to keep damages and the associated down-times for repair or exchange relatively low. Some of these elements are collected in the following list, but not exhaustively. Drives (motors, bearings and seals) can be effectively protected if drives are irreversibly disabled on low lubricant or coolant level and if changes of speed are restricted to relatively small in/decrements. Emergency stop of sterilization sequence(s) is easily activated in cases such as pressure drop or stirrer failure. SOPs for reactor preparation are advantageously implemented for on-line use on a process computer, e.g., for calibration, checking (of zero and slope) including basic trouble shooting of in-situ sensors. Re-calibration of analytical instruments (e.g., exhaust gas analyzers, once per hour) is repeatedly activated. Disabling (freezing) of controllers is essential if there are indications that the sensor might not measure representatively (e.g., pH-probe not submerged during draw and fill cycle). Infections are minimized if the reactor and supporting vessels are held under a small overpressure at any time they are in use, etc. At first glance these hints may look like a waste of time and money but, in the end, they help effectively to minimize unproductive time for the equipment – most important for scientific research – to reduce the probability of producing bad results.

5 Non-Invasive Techniques

If a proper sampling unit and an appropriate sample pretreatment/inactivation method is applied, a variety of routinely used standard instruments such as gas and liquid chromatographs (GC and LC), flow injection (FIA) or field flow fractionation (FFF) analyzers, capillary electrophoreses (CE) or mass spectrometers (MS) can be exploited for improved analyses of bioprocesses. However, it is mandatory to include automated sampling and preparation in the respective instrumental analysis equipment intelligently! This approach avoids both badly reproducible sampling artefacts, which are generally difficult to recognize, and, due to the very small sample streams needed, avoids relevant significant disturbances of bioprocesses as well. It is practically "non-invasive" and, hence, optimally suited to qualify and quantify growing and/or producing populations in vivo.

5.1 Well Defined and Precisely Known Operating Conditions

The most versatile approach among those referred to is the FIA because it allows one to benefit from the know-how, skill and experience globally accumulated over decades in the entire field of wet and instrumental chemistry. It is used to work outside the monoseptic space and needs an appropriate interface to the process. Dilutions can be made by a factor of 1000 within less than 1 min and a precision of better than $\pm 1\%$ which is a figure far beyond the capacity of manual procedures [58]. Nearly any chemical reaction can be realized and combined with physical methods like diffusion or extraction. Most important, however, is that all these options can be easily combined to an even complicated analytical protocol which can be executed automatically. Presently, the detection of a selected product after being produced or separated using an FIA is achieved by relatively simple detectors such as optical, fluorimetric, physical, electrochemical or even bio-sensors. In cases of whole cells, cell compartments or aggregates, and even macro-molecules, it certainly would be desirable to implement classifying detection principles such as mass spectrometry or flow cytometry in order to gain the important (but presently not exploited) information about an analyte's population distribution. For the time being, this has not yet been realized.

Dynamic transitions or responses of populations in bioprocesses can be pin-pointed and quantitatively characterized. Proliferation and product formation of microbes and cell populations are highly dynamic processes. Efficiency of growth and product formation depends on physiological state or transitions between distinct physiological states which must be properly tuned. It is, therefore, essential to know the (most) relevant dynamic responses of biosystems to macro- and micro-environmental factors in order to design straightforward bioprocesses optimally and to control them successfully. In many critical situations (e.g., during fed batch or continuous cultivation) only automated sampling and analysis will be able to cope with the short relaxation times of the culture (see Fig. 3). A robust process design can be carried out reliably only if several key factors are qualitatively and quantitatively identified: the relevant (physiological) effectors with their dominating relaxation time regimes and the dynamics of transitions from one physiological state to another.

5.2 Disturbances of the Cultures – Sensitivity

The so-called culture parameters must be precisely supervised. Even very short deviations of any intentional parameter from the desired setpoint must be trapped and documented. It is also a necessary requirement for tracing back and analyzing any unexpected process behavior since

1. even small deviations from setpoint may cause considerable variations of chemical and biological state variables and
2. return to the expected (steady) state may be quite delayed (see, e.g., Fig. 7 in [33]).

The responses of biosystems can, in our experience, sometimes be a better indicator of the constancy of an operational variable than direct measurement. This has been found, for instance, in continuously grown synchronized yeast cultures: an interruption of the medium feed of the order of only 5 s caused a significant change in the period of the oscillation affected (5 to 10 min [59]). The other way round creates responses as well: Münch er al. [31] have shown that pulses of as little as $60 \, \text{mg} \, l^{-1}$ of glucose to a continuous yeast culture operated with $30 \, \text{g} \, l^{-1}$ glucose feed at an intermediate dilution rate $(0.08 < D < 0.20 \, \text{h}^{-1})$ are sufficient to affect the frequency of the synchronous oscillation. If the pulses are regularly repeated, the oscillation period can be stably controlled within a permissive window (e.g., shortened to 60% of the spontaneous value) which means that the generation time is affected and can be tuned by minute operational changes of the microenvironment.

5.3 Oscillations

Oscillatory cultures are usually neither well understood nor desirable. There is, however, one important aspect in scientific research, namely synchronized cultures, which permit study of cell cycle dependent events and relations under well controlled and reproducible cultivation conditions. Yeasts of the type *Saccharomyces* are an outstanding example since they provide an opportunity to run pertinent, undampened and stably synchronized populations [31, 60]. Such a system is ideal for studying, for instance, mechanisms of self-synchronization and cell-signalling.

The special tendency of these yeasts to synchronize spontaneously is based on the mobility of the specific storage carbohydrate content of cells which is one determinant characteristic property governing progression through the cell cycle. Storage materials such as trehalose and glycogen are mobilized during the S-phase and built-up during the G1-phase. It is thought that a critical upper level of storage material content triggers cells to enter into the S-phase autonomously. Cells having accumulated a high but not yet critical level of storage material content – in a permissive window right below this critical upper level – however, are susceptible to external triggers such as extra carbon and/or energy supply. As a synchronous population (or subpopulation) develops with time its distribution broadens (i.e., the variance increases). This would end up in an ideally asynchronous population distribution if there were not a re-synchronization signal. In our comprehension, this signal is generated by the few cells in the very leading head of the (broadening) distribution: as they reach the upper limit of specific storage carbohydrate content they enter the S-phase spontaneously, mobilize their storage carbohydrates during this phase rapidly and necessarily form and excrete ethanol (and other metabolites to a smaller extent). This ethanol, in turn, signals and triggers the fraction of cells lying in the permissive window which immediately respond by entering the S-phase too, thus amplifying the signal. The result is a sudden decrease of the variance of the distribution of this subpopulation, i.e., a re-synchronization. This effect is rapid and efficient enough to explain the quick occurrence of synchronous oscillations when a continuous culture is shifted to permissive conditions (i.e., basically $0.1 \, h^{-1} \leq D \leq 0.22 \, h^{-1}$ at pH 5.0 and 30.0°C under glucose limitation).

As opposed to the introductory statement, oscillatory cultures might be advantageous, namely whenever a non-oscillating culture suffers from intracellular limitation which can, principally, be relaxed by an extracellular challenge (as discussed above). Neubauer et al. recently came to an equivalent conclusion [61] when they investigated the performance of an *E. coli* culture in a cyclic CSTR-PFR-system under various limitation conditions. Whenever the cells were repeatedly challenged by a (relatively) high glucose consumption (in the PFR-part of the system), the overflow metabolite acetate was formed there and also excreted (because of an intracellular limitation) but it did not accumulate because it was effectively consumed in the CSTR-part. However, when the extracellular oxygen limitation was responsible for acetate formation, the overall biomass yield decreased and acetate formation was high. The authors speculated that these artificially created short-term heterogeneities influence the physiology and may be of major importance for process performance. This nicely confirms our view that intracellularly limited cultures are easily disturbed and relatively unstable. The consequence for industrial applications, where only non-ideal reactors can be used, is that the large-scale process may be significantly more robust because the oscillatory movement through different micro-environments permanently challenges the cells and keeps them safely off an intracellular limitation.

5.4 Energy Metabolism (Energy Charge, Reducing Equivalents)

The energy metabolism or energetic state of cells can be quantified by determining the actual concentrations of phosphorylated nucleotides and of the carriers of reducing equivalents. The concentrations and their ratios of these components are determinant regulators of metabolism. The respective analyses are quite tricky since the individual component-levels of these energy currency units have a very short life-time, namely of the order of seconds [5, 62, 63]. The conclusion from this knowledge is to best avoid sampling – if an excellent technique is not available (as described by [5]) – followed by chemical reactions, but rather measure non-invasively either directly or estimate from indirect measurements. Of course, this is easier said than done. One must then rely on magnetic, electrical or optical measuring principles. They are all available yet they have their disadvantages as well.

Culture fluorescence, for instance, allows one to determine the pool of the reduced form of the nicotinamide adenine dinucleotides (NADH and NADPH) but not of the respective oxidized forms. Consequently, the signal yields relative information only ("relative intracellular reduction degree"). This is also why the signal can – with tremendous caution – be used to estimate biomass concentration but this can work only if the ratio of reduced/oxidized forms of the pool components remains constant over the time period of observation.

Redox potential signals a lumped reduction degree of the entire culture liquid which is a complicated superposition of the concentrations and qualities of many unknown abiotic and cell-associated components. Therefore, its use to interpret the cell's energy metabolism is quite limited, and it serves as a simple rough indicator rather than a reliable direct measure. It does not, in principle, deliver information on the intracellular redox state of cells.

Nuclear magnetic resonance (NMR) spectroscopy is an outstanding tool to acquire, non-invasively, quantitative information about the energy currency units (and many others) in a cell. A great part of this issue is dedicated to NMR techniques and, therefore, just some caveats are mentioned in this context: the method is inherently insensitive and requires either very highly concentrated analytes ($> 10^{10}$ cells per ml) or long measuring times (minutes to hours) or both.

Last but not least, there is yet one other methodology to determine the energetics of growth and product formation indirectly, non-invasively and accurately: heat flux calorimetry. The heat generated during bioactivity depends on the stoichiometry of the (lumped) bioreactions as well as on the nature of the reactands, i.e., the general degree of reduction and molecular size of the reactands. Chemical reactions are driven by a change in free energy which consists of an enthalpic and an entropic part. It is the enthalpic fraction which is determined by (heat flux) calorimetry [64] and there are powerful tools at hand to correlate the measurements with thermodynamic efficiencies of growth and product formation [62, 65–67]. The method is so potent that even slow growing organisms such as hybridoma cells [68] or microbes with a very low biomass yield such as methanogens [69] can be effectively studied.

5.5 Behavior in High Density Cultures

The volumetric conversion rates in high density cultures are very high, indeed. This fact necessitates the use of rapid on-line analysis as well as the full exploitation of additional calibration and mathematical compensation methods (as discussed above). Off-line analytical methods are most likely to fail due to immense errors introduced by (manual) sampling. For instance, Nipkow et al. [70] reported that a *Zymomonas mobilis* culture at nearly $100 \, g \, l^{-1}$ dry weight produced CO_2 at a volumetric rate equivalent to a 3 vvm aeration resulting in highly foamy samples from which no exactly quantitated volume-aliquot could be taken for further analyses.

It is very likely that high concentrations of metabolites occur in high density cultures. These metabolites may exert significant inhibition on growth and/or product formation and this is why they should be monitored and, eventually, controlled. But this is not to be discussed here. Moreover, spatial limitation of the cells may occur as well and cells may behave differently in more dilute culture systems due to cell-cell interactions, diffusion controlled reactand transfer or osmotic problems. For simplicity, let us assume cells to be spherical and rigid; then the subspace which can be occupied by those cells is two thirds of the available space, e.g., the working reactor volume; one third is void volume; the cellular specific density to be approximately $1000 \, g \, l^{-1}$ (wet weight per cell volume) and the dry matter fraction of cells to be 20%. Then the maximal attainable cell density is estimated to be of the order of $0.2 \, 1000 \, 0.67 \approx 130 \, g \, l^{-1}$ (dry weight). This is well of the order observed in experimental studies [71, 72]. Contrary to those findings, Goma [73,74] showed that, with yeast cell recycle cultivations, cell dry mass concentrations of greater than $300 \, g \, l^{-1}$ could be stably attained. His explanation was a dramatic change (i.e., a decrease by 50%) of the water content of the cells during such modes of operation. Another likely explanation, however, is that the analytical methods presently available for biomass estimation are not usable and are erroneous for high cell density systems. Although most probably not of interest for industrial processes (for various reasons) these very high cell density systems may be of considerable importance in physiological research if the appropriate analytical tools are provided (see other contributions on NMR-analyses in this issue).

5.6 Cellular Activities

Cellular activities such as those of enzymes, DNA, RNA and other components are the primary variables determining the performance of microbial or cellular cultures. The development of specific probes for measurement of these activities in vivo is therefore of essential importance in order to get direct analytical access to these primary variables. What we can measure today are secondary variables such as the concentrations of metabolites which fully depend on primary and operating variables.

There are many highly specific analytical methods proposed (as reviewed, e.g., by [75]) covering a variety of extracellular or periplasmatically located components on-line but the successes for those suited for determination of intracellular activities are virtually nil. One interesting approach was the indirect analysis of viability (or death rate parameter) of CHO-cells by on-line analysis of lactate dehydrogenase activity found in the culture supernatant [76, 77].

RNA content, although extremely important in governing the expression level of enzymes, is not yet accessible on-line though this is realistically imaginable. This decisive cellular compartment has a wide dynamic range depending on the specific growth rate, it varies with the supplied substrate and must be considered in structured models of microbes and cells [27, 28, 78].

It is believed there is only one report describing a virtually non-invasive, fully automated method for the determination of (relative) DNA content [79, 80]. This methods works continuously in a flow analysis (i.e., a FIA without injection) as culture liquid is steadily removed, diluted, fixed, stained and the fluorescence of the DNA-bound stain quantitated. Unfortunately, this method can estimate only the lumped DNA content as it was run with an ordinary fluorospectrophotometer. A flow cytometer would, eventually, allow one to resolve the population distribution as well. Obviously, on-line flow cytometry has never been realized or even tried in bioprocess monitoring but would be extremely helpful for verifying highly dynamic segregated models.

6 Conclusions

The undelayed evaluation of a culture's state by using software sensors and computers, based on the quantitative analytical information provided by hardware sensors and intelligent analytical subsystems, constitutes an excellent basis for targeted process control. Experts – either humans or computers – have the data and the deterministic knowledge to trace observed behavior back to the physical, chemical and physiological roots, thereby gaining a qualitative improvement of bioprocess control, a quantum leap: process control can act on the causes of effects rather than just cure symptoms. A simple SOP (standard operating procedure) has proven useful, namely

1. measure everything that can be measured in the very beginning of process development,
2. decide whether or not a variable is relevant,
3. choose the relevant variables to be controlled and/or documented,
4. collect all raw data at any time and distinguish on-line between variable and parameter behavior, organize an archive of all these data accordingly and don't discard seemingly useless data since they contribute to the treasure of experience.

If it is correct that today's bioengineering, with all its tools and methodologies, is too slow and not efficient enough, then it is all the more urgent to improve the performance of the methods, tools and equipments yet available and invent new and better ones. In essence, techniques of operation, reproducibility and casual-analytical interpretation of observations and measurements need massive impulses.

Acknowledgement. The financial support of this work through Swiss Priority Program in Biotechnology is gratefully acknowledged.

7 References

1. Villadsen J (1995) ECB7, Nice, F (Feb 1995): MAC161
2. Fiechter A, Fuhrmann GF, Käppeli O (1981) Adv Microb Physiol 22: 123
3. Sonnleitner B, Hahnemann U (1994) J Biotechnol 38: 63
4. Nielsen J, Villadsen J (1993). In: Rehm HJ, Reed G, Pühler A, Stadler P, Stephanopoulos G (eds) Biotechnology, 2nd ed, VCH Weinheim
5. Rizzi M, Theobald U, Baltes M, Reuss M (1993). In: Nienow AW (ed), Bioreactor and bioprocess fluid dynamics, Mechanical Engineering Publ Ltd, London
6. Lübbert A (1992) J Biotechnol 25: 145
7. Meyer HP (1992; guest editor of special issue) J Biotechnol 22(1, 2)
8. Sonnleitner B (1992; guest editor of special issue) J Biotechnol 25(1, 2)
9. Markx GH, Davey CL, Kell DB (1991) J Gen Microbiol 137: 735
10. Nakajima M, Siimes T, Yada H, Asama H, Nagamune T, Linko P, Endo I (1992) Biochemical Engineering for 2001 (Furusaki S, Endo I, and Matsuno R, eds) Springer Verlag Tokyo: 681
11. Kell DB, Sonnleitner B (1995) TIBTECH: 13: 481
12. Filippini C, Moser JU, Sonnleitner B, Fiechter A (1991) Anal Chim Acta 255: 91
13. Benthin S, Nielsen J, Villadsen J (1992) Biotechnol Bioeng 40: 137
14. Rothen SA, Saner M, Meenakshisundaram S, Sonnleitner B, Fiechter A (1995) J Biotechnol (submitted)
15. Heinzle E, Oeggerli A, Dettwiler B (1990) Anal Chim Acta 238: 101
16. Locher G, Hahnemann U, Sonnleitner B, Fiechter A (1993) J Biotechnol 29(1/2) : 75
17. Royce PN (1992) Biotechnol Bioeng 40: 1129
18. Noorman HJ, Luijx GCA, Luyben KCAM, Heijnen JJ (1992) Biotechnol Bioeng 39: 1069
19. Ponti C (1993) PhD thesis, ETH # 10504, Zürich
20. Konstantinov KB, Yoshida T (1992) J Biotechnol 24: 33
21. Locher G, Sonnleitner B, Fiechter A (1990) Bioproc Eng 5: 181
22. Montague G, Morris J (1994) TIBTECH 12: 312
23. Posten C, Gollmer K (1995) ECB7, Nice, F; Abstract book Vol III: MEP-147

24. Simutis R, Havlik I, Lübbert A (1993) J Biotechnol 27: 203
25. Stephanopoulos G, Locher G, Duff M (1995) CAB6, Garmisch-Partenkirchen, D, preprints: 195
26. Ohashi E, Karube I (1995) J Biotechnol 40(1) : 13
27. Benthin S, Nielsen J, Villadsen J (1991) Biotechnol Tech 5(1) : 39
28. Nielsen J, Pedersen AG, Strudsholm K, Villadsen J (1991) Biotechnol Bioeng 37: 802
29. Locher G, Sonnleitner B, Fiechter A (1991) J Biotechnol 19: 173
30. Locher G, Hahnemann U, Sonnleitner B, Fiechter A (1993) J Biotechnol 29(1/2) : 57
31. Münch T, Sonnleitner B, Fiechter A (1992) J Biotechnol 24: 299
32. Locher G, Sonnleitner B, Fiechter A (1992) J Biotechnol 25: 55
33. Fiechter A, Sonnleitner B (1994) Adv Microb Physiol 36: 145
34. Weuster-Botz D, Wandrey C (1995) Process Biochem 30: 435
35. Sonnleitner B, Bomio M (1990) Biodegradation 1: 133
36. Bradley J, Kidd AJ, Anderson PA, Dear AM, Ashby RE, Turner APF (1989) Analyst 114: 375
37. Locher G, Sonnleitner B, Fiechter A (1992) Proc Contr Qual 2: 257
38. Filippini C, Sonnleitner B, Fiechter A, Bradley J, Schmid R (1991) J Biotechnol 18: 153
39. Nipkow A, Andretta C, Käppeli O (1990) Chem Ing Tech 62: 1052
40. Kell DB, Markx GH, Davey CL, Todd RW (1990) Trends Anal Chem 9: 190
41. Benthin S, Nielsen J, Villadsen J (1991) Anal Chim Acta 247: 45
42. Thatipamala R, Rohani S, Hill GA (1991) Biotechnol Bioeng 38: 1007
43. Thatipamala R, Rohani S, Hill GA (1994) J Biotechnol 38: 33
44. Fehrenbach R, Comberbach M, Pêtre JO (1992) J Biotechnol 23: 303
45. Konstantinov KB, Pambayun R, Matanguihan R, Yoshida T, Perusich CM, Hu WS (1992) Biotechnol Bioeng 40: 1337
46. Packer HL, Keshavarzmoore E, Lilly MD, Thomas CR (1992) Biotechnol Bioeng 39: 384
47. Pons MN, Wagner A, Vivier H, Marc A (1992) Biotechnol Bioeng 40: 187
48. Sonnleitner B, Locher G, Fiechter A (1992) J Biotechnol 25: 5
49. Tanaka H, Aoyagi H, Jitsufuchi T (1992) J Ferment Bioeng 73: 130
50. Hibino W, Kadotani Y, Kominami M, Yamane T (1993) J Ferment Bioeng 75: 443
51. Li RC, Nix DE, Schentag JJ (1993) Antimicrob Agents Chemotherapy 37: 371
52. Shiloach J, Bahar S (1993) ECB6, Firenze, I; Proceedings Vol 2: TU156
53. Bedard C, Jolicoeur M, Jardin B, Tom R, Perret S, Kamen A (1994) Biotechnol Technique 8(9): 605
54. Nielsen J, Johansen CL, Villadsen J (1994) J Biotechnol 38: 51
55. Markx GH, Kell DB (1995) Biotechnol Prog 11: 64
56. Suhr H, Wehnert G, Schneider K, Bittner C, Scholz T, Geissler P, Jähne B, Scheper T (1995) Biotechnol Bioeng 47: 106
57. Wu P, Ozturk SS, Blackie JD, Thrift JC, Figueroa C, Naveh D (1995) Biotechnol Bioeng 45: 495
58. Garn MB, Gisin M, Gross H, King P, Schmidt W, Thommen C (1988) Anal Chim Acta 207: 225
59. Sonnleitner B, Fiechter A (1988) Anal Chim Acta 213: 199
60. Strässle C, Sonnleitner B, Fiechter A (1989) Biotechnol 9: 191
61. Neubauer P, Häggström L, Enfors SO (1995) Biotechnol Bioeng 47: 139
62. Larsson C, von Stockar U, Marison I, Gustafsson L (1995) Thermochimica Acta 251: 99
63. Zeng AP, Deckwer WD (1995) Biotechnol Prog 11: 71
64. Birou B, von Stockar U (1989) Enz Microb Technol 11: 12
65. Sandler SI, Orbey H (1991) Biotechnol Bioeng 38: 697
66. Heijnen JJ, van Dijken JP (1992) Biotechnol Bioeng 39: 833
67. von Stockar U, Larsson C, Marison IW, Cooney MJ (1995) Thermochim Acta 250: 247
68. Randolph TW, Marison IW, Berney C, von Stockar U (1989) Biotechnol Tech 3: 369
69. Schill N, von Stockar U (1995) Thermochim Acta 251: 71
70. Nipkow A, Sonnleitner B, Fiechter A (1986) J Biotechnol 4(1) : 49
71. Riesenberg D (1991) Curr Opin Biotechnol 2: 380
72. Korz DJ, Rinas, U, Hellmuth K, Sanders EA, Deckwer WD (1995) J Biotechnol 39: 59
73. Goma G, Durand G (1988) Proceedings of 8th International Biotechnology Symposium (Paris, 1988; Durand G, Bobichon L, Florent J, eds) Vol 1: 410
74. Goma G (1994) ESF-workshop on "Intensification of Biotechnological Processes", March 94, Davos, CH
75. Degelau A, Freitag R, Linz F, Middendorf C, Scheper T, Bley T, Müller S, Stoll P, Reardon KF (1992) J Biotechnol 25: 115
76. Renner W, Jordan M, Eppenberger HM, Leist C (1993) Biotechnol Bioeng 41: 188

77. Biospectra AG (1995) Company pamphlet of Biospectra AG, Zürcherstr 137, CH-8952 Schlieren
78. Cortessa S, Aon JC, Aon MA (1995) Biotechnol Bioeng 47: 193
79. Münch T, Rothen SA, Sonnleitner B, Fiechtner A (1993) ECB6, Firenze, I; Proceedings Vol 2: TU384
80. Sonnleitner B (1993) Bioreactor Performance (Mortensen U, Noorman HJ, eds) Biotechnology Research Foundation, Lund: 143
81. Rothen SA, Sonnleitner B, Witholt B (1995) ECB7, Nice, F; Abstract book Vol III: MEP-101

Author Index Volume 54

Subject Index

Springer-Verlag
and the Environment

We at Springer-Verlag firmly believe that an international science publisher has a special obligation to the environment, and our corporate policies consistently reflect this conviction.

We also expect our business partners – paper mills, printers, packaging manufacturers, etc. – to commit themselves to using environmentally friendly materials and production processes.

The paper in this book is made from low- or no-chlorine pulp and is acid free, in conformance with international standards for paper permanency.